The Queen of the Sciences: A History of Mathematics

Part I

Professor David M. Bressoud

PUBLISHED BY:

THE TEACHING COMPANY
4151 Lafayette Center Drive, Suite 100
Chantilly, Virginia 20151-1232
1-800-TEACH-12
Fax—703-378-3819
www.teach12.com

Printed in the United States of America

ISBN 1-59803-427-8

David M. Bressoud, Ph.D.

DeWitt Wallace Professor of Mathematics
Macalester College, St. Paul, MN

David Bressoud earned a Bachelor's degree in Mathematics from Swarthmore College in 1971; joined the Peace Corps, where he taught mathematics and science at the Clare Hall School in Antigua, West Indies; and then earned a Ph.D. in Mathematics at Temple University in 1977. He taught for 17 years at the University Park campus of The Pennsylvania State University before moving to Macalester College in 1994, where he chaired the Department of Mathematics and Computer Science from 1995 until 2001. He has received a Sloan Foundation Fellowship and a Fulbright award and has held visiting positions at the Institute for Advanced Study, the University of Wisconsin–Madison, the University of Minnesota, Université Louis Pasteur (Strasbourg, France), and State College Area High School. His research interests lie in number theory, combinatorics, and analysis, and he has published more than 50 research papers.

Professor Bressoud's interest in the history of mathematics was sparked by a series of lectures by Subrahmanyan Chandrasekhar on Newton's *Principia* and by his own research into the mathematics of the self-taught genius Srinivasa Ramanujan. He has since taught courses on the history of mathematics in South Asia and on Newton's *Principia* and the scientific revolution.

In addition to two textbooks on number theory, *Factorization and Primality Testing* (1989) and *A Course in Computational Number Theory* (2000; coauthored with Stan Wagon), Professor Bressoud has written four textbooks that draw on the history of mathematics to motivate its study: *Second Year Calculus from Celestial Mechanics to Special Relativity* (1991); *Proofs and Confirmations: The Story of the Alternating Sign Matrix Conjecture* (1999); *A Radical Approach to Real Analysis* (2nd edition, 2007); and *A Radical Approach to Lebesgue's Theory of Integration* (2008). In 1994 he won the Award for Distinguished College or University Teaching of Mathematics from the Allegheny Mountain Section of the Mathematical Association of America (MAA). He has also received the MAA's Beckenbach Book Prize for *Proofs and Confirmations* (2000) and Macalester College's Thomas Jefferson Award for his "personal influence, teaching, writing and scholarship" (2005). Professor Bressoud has been elected president of the Mathematical Association of America for 2009–2010.

Acknowledgments

I want to thank Nancy Eskridge for all of her assistance as I put this course together. She helped me sharpen my focus, and she always had a good sense of what would and would not work. I also want to express my appreciation to the rest of my team at The Teaching Company, especially Nelson Ginebra, who has done a great job overseeing production; Lisa Robertson, who tracked down many of the images; Marcy McDonald; and Zach Rhoades. I was always impressed by both their professionalism and the sheer enjoyment they share in what they do. I owe a debt to my wife, Jan, as well as the many friends and alumni of Macalester College who rose early on Wednesday mornings in the fall of 2007 to hear the first drafts of these lectures and who gave me many good suggestions for improvements.

Table of Contents

The Queen of the Sciences: A History of Mathematics
Part I

The Queen of the Sciences: A History of Mathematics

Scope:

These lectures describe the historical development of mathematics from the earliest records of Mesopotamia and Egypt up to the problems and challenges of mathematics today. The goal is to convey the nature and power of this discipline while exploring the lives of many of the people who have shaped it. Although our focus is on the great figures, we will also study the men—and occasional women—who laid the groundwork for their discoveries, acted as their collaborators, and built on their insights.

The history of mathematics covers 4000 years and many different cultures and civilizations. These lectures have been structured so that developments of the 17th century in Europe form the pivot. That is the critical time and place in which developments in algebra, geometry, astronomy, mechanics, and the mathematics of motion reach the level of sophistication needed to weave them into a structure that would enable the rapid development of mathematics in the succeeding centuries. The 17th century witnessed the emergence of analytic geometry and calculus as well as important foundational advances in geometry, probability, and number theory.

Lectures Ten through Fourteen are devoted to the mathematical advances of the 17th century, focusing on seven great people: Galileo, a modern figure forever constrained by Aristotelian assumptions; Napier, the Scotsman who would use logical reasoning to simplify complex calculations; Fermat, the parliamentary lawyer who would establish the foundations of calculus, set number theory on a new course, and discover a problem that would challenge the greatest minds of the 20th century; Newton, whom Keynes described as the "last of the magicians"; Leibniz, the court librarian who would fight with Newton for bragging rights to the creation of calculus; and the Bernoullis, two incredibly gifted Swiss brothers who would become Leibniz's collaborators.

The lectures leading up to the 17th century explore the question of how mathematics arrived at this golden moment. We begin by considering the definition of mathematics and the question of why it has been so unreasonably effective at explaining the physical world. Our exploration of the historical record begins with Babylonian and

Egyptian mathematics of 2000 B.C. Already, this is very sophisticated mathematics that uses the Pythagorean theorem, solves quadratic equations, and calculates square roots to high precision. We will spend three lectures on Greek and Hellenistic mathematics. The first will follow its development through Thales, Pythagoras, Theaetetus, and Eudoxus up to the crowning achievement of Euclid's *Elements*. In the next lecture, we will focus on three great Hellenistic figures: Archimedes, Apollonius, and Diophantus.

We will conclude our study of this period with an exploration of the early development of trigonometry and the contributions of Hipparchus and Ptolemy. We continue with the Indian, Chinese, and Islamic mathematics that continued this chain of development toward the 17th century, including computational advances from China, trigonometry from India, and algebra from the Islamic caliphates. The final preparation would come from the European algebraists, focusing particularly on Tartaglia and Cardano, the feuding Italian algebraists of the 16th century.

Following the 17th century, this survey of mathematical history is necessarily much sketchier. The creative power and complexity of the mathematics of the past 300 years is so great that we can no more than sample a few of its results. I have chosen the people and mathematical developments that I consider most interesting, among them Euler, the man who dominated 18th-century mathematics even after he had gone blind; and Gauss, the "prince of mathematics," whose reluctance to publish until a result was perfect would frustrate many of his contemporaries. Several of these mathematicians of the past 300 years died tragically young: Galois at 20, Abel at 26, Ramanujan at 32, and Riemann at 39. Several, such as Jacob Jacobi, James Joseph Sylvester, Sophie Germain, and Sofya (Sonya) Kovalevskaya, had to fight prejudice against Jews or women.

In these last lectures, we will investigate what happened to calculus in the 19th century as it came to be known as *analysis*. We will look at developments in geometry and algebra as both began to focus on invariants, the fundamental characteristics that do not change. We also will explore the development of elliptic and modular functions that arose in extending the notion of trigonometric functions and that would prove crucial to modern physics. We will follow some of the recent developments in number theory, including the proof of Fermat's last theorem. We will examine the role of mathematics in predicting physical reality, as exhibited in the work of Maxwell and Einstein. And we will conclude with a selective survey of current problems and areas of study in mathematics.

Lecture One

What Is Mathematics?

Scope:

This lecture explores the nature of mathematics, a subject that arose from the abstraction of patterns observed in the world around us and developed as those abstractions were codified and pushed beyond practical applications. One of the mysteries of mathematics is its unreasonable effectiveness: Why it is that abstractions that arose in one context can lead to unexpected insights when applied to a totally different situation? We will explore how the history of mathematics can help us understand the true nature and power of this subject. The lecture concludes by outlining the path this course will take over 4000 years in the history of mathematics.

Outline

I. Mathematics is about ideas, and one of the most effective ways of understanding mathematical ideas is to see how they developed. Where did they come from? What were people trying to understand as they looked at these ideas? How did the ideas arise from different civilizations and cultures?

A. In exploring mathematical ideas, it's also important to know where difficulties arose in their development. As we'll see, great mathematicians have struggled with some of the ideas we'll explore; thus, it's no surprise that students may struggle with them as well.

B. The history of mathematics is also full of great stories and interesting people. In this course, we'll meet such figures as Leonhard Euler, a great mathematician who ultimately went blind; Evariste Galois, who solved one of the most fundamental problems in mathematics at the age of 17; and the self-taught Indian mathematician Srinivasa Ramanujan.

II. Before we launch into the history of mathematics, we should define the field.

A. The essence of mathematics is the abstraction of pattern. Mathematicians pick out patterns from the world around us and then abstract those patterns in order to manipulate them and tell us more about the world.

B. Some of the simplest examples of this abstraction of pattern come from numbers. For example, we consider number simply as a qualifier: 5 people, 5 stars. From that, we can abstract the number 5. That abstraction is useful as we begin to combine objects; it becomes something that goes beyond the objects themselves.

C. The same thing happens in geometry. The basic geometric objects, such as lines and circles, express spatial relationships. We might think of a tree at 1 point, a house at another point, and a person standing in between them. We consider these 3 points and abstract them into a triangle.

D. What is the reality behind these abstractions? Is 5 something that actually exists? Does a pure Platonic circle really exist? The adjective *Platonic* reminds us of Plato's view that these ideal abstractions do exist, that there is a reality beyond ourselves that we tap into as we make these abstractions.

E. This leads to an even more fundamental question: Was mathematics discovered or created? I believe that as we develop these abstractions of patterns, we are trying to explain a deeper reality that does exist. Most mathematicians and scientists realize that we can never truly describe this reality, but mathematics creates a language, a set of symbols, that enables us to work with aspects of this deeper reality.

F. Mathematicians look for points of similarity between the patterns they discover. These points of similarity, in turn, suggest ways in which the pattern might be extended. This kind of exploration, the search for new knowledge of the world, is what makes mathematics exciting.

III. The patterns of mathematics come from many sources, as we'll see throughout this series of lectures.

A. One of the most important sources is commerce and civil administration.

1. This source gives us basic arithmetic—the fundamental operations of addition, subtraction, multiplication, and division—and simple fractions, as well as the whole numbers, or integers.

2. Commerce and civil administration also give rise to rates. We may need to know, for example, how long it will take to build a wall; what is the rate at which this job can be accomplished? We may need to know the yield for a given plot of land; how much is produced on each acre?
3. Commerce and civil administration are also the origins of much work in algebra, which in its earliest forms is closely connected to geometry.

B. Another source of the patterns in mathematics is navigation and surveying.

1. The need to make land measurements gave rise to geometry, which literally means "measuring the earth." Measurements of distances, areas, and volumes were needed.
2. As we'll see, geometry motivated work in algebra and number theory. Number theory relates to understanding the structure of integers. How can the integer 20 be represented? Can we write it as the sum of two squares? How can we decompose 20 into a product of primes?

C. A third source of the patterns of mathematics is astronomy or astrology.

1. The ancients made no clear distinction between these two fields. They studied the heavens to try to understand what was likely to happen on earth.
2. Some of the greatest astronomers, including Johannes Kepler, were also astrologers. Kepler's work in both astronomy and astrology would lay the foundations for much of the development of calculus.
3. This lack of distinction between astronomers and astrologers carried over to mathematicians. The emperor Tiberius is said to have banished all "mathematicians" from Rome. In fact he banished the astrologers, who were predicting his downfall.
4. Some of this confusion stems from the fact that important advances in mathematics came directly out of astronomy. By looking at the heavens, mathematicians were able to pick out patterns in much purer form than they could in the world around them.

5. We'll also see mathematics arising from other physical phenomena such as optics, electricity, and magnetism, and even the study of subatomic particles. Much mathematics has been created from the search for explanations of physical phenomena.

D. Yet another source of mathematical patterns is art and architecture.

1. In later lectures, we'll see how the ideas of symmetry would come to play a fundamental role in the mathematics developed in the 19th and 20th centuries.
2. The Lion Court at Alhambra in Granada is a wonderful example of symmetry in art; through this kind of symmetry, many mathematicians have seen the kinds of patterns they could apply to other problems, both in geometry and algebra.

IV. Let's look further at number and distance.

A. At some point in antiquity, someone decided to apply the idea of number to distance, but there is no natural unit for measuring distances. Feet or meters, for example, are human constructs. Even once we decide on a unit, that unit is not necessarily appropriate for the distance we're trying to measure.

B. Let's say that we're going to measure a distance using a stick. Almost inevitably, as we get close to the endpoint, we find that the distance left is less than the length of the stick. One way to solve this problem is to mark the stick off into smaller intervals.

C. Again, however, the smaller pieces may not measure the remaining distance exactly. We can subdivide the smaller pieces, but we begin to realize that this process will be never ending. Does it even make sense to talk about the distance between two points as a number if we can never express that number exactly? Babylonian and Greek mathematicians wrestled with this question.

D. Trying to apply number to time introduces other problems. Unlike distance, for which there is no natural unit, for time, there are too many units. Time involves the day, the lunar month, and the solar year, each of which is incommensurable with the other. We cannot measure a lunar month in a precise number of days. We cannot measure a solar year in an exact number of lunar months or an exact number of days. Whichever of these units we use, we will have a little bit left over.

E. Further, these different units are traveling in different cycles; this would present fascinating problems for mathematicians, such as how to determine when all these cycles would line up or when different planets would be in alignment.

V. Two famous papers will be useful in our exploration of the nature of mathematics.

A. The first of these was published in 1960 by the physicist Eugene Wigner, who taught at Princeton. Wigner's paper is called "The Unreasonable Effectiveness of Mathematics in the Natural Sciences."

1. Wigner noted that physicists seek to understand patterns in the world around them; they then look to mathematics to see if they can find similar patterns. Can the patterns in mathematics be applied to physical phenomena?
2. Wigner remarked that in this comparison of patterns in the natural world and patterns from mathematics, the mathematical pattern often fits the physical phenomenon with amazing accuracy. Further, the mathematical pattern frequently offers excess content. It is possible to use the mathematical model to see beyond the original phenomenon.
3. In other words, not only does the mathematical model enable us to make predictions about what we've already seen, but it also suggests a deeper reality. Mathematics represents the ultimate physical reality, showing us things we never would have expected without the mathematical models.
4. Given that mathematicians get their ideas from the world around them, it's perhaps not surprising that mathematics fits physical phenomena. It may be surprising, however, that patterns pulled from one area apply to a totally unrelated area.
5. As we'll see, the study of astronomical phenomena gave rise to trigonometric functions, but these trigonometric

functions can also be applied to problems in mechanics, heat flow, and electricity and magnetism.

B. The second paper, "The Unreasonable Effectiveness of Mathematics," was written by Richard Hamming in 1980 as a response to Wigner's article. Hamming was a prominent applied mathematician working in signal processing at Bell Labs.

1. Hamming's article does a good job of clarifying the definition of mathematics. He describes the four faces of mathematics, two of which we've already talked about: number and geometry.

2. The third face of mathematics is close reasoning. This tool helps mathematicians keep track of the multitude of patterns and abstractions they work with and reach legitimate conclusions. Formal logic, from the Greek tradition, is one way of expressing close reasoning.

3. The fourth face of mathematics is artistic taste. As we look for patterns, we find that there is an element of aesthetic appeal to good mathematics.

VI. These lectures will cover 4000 years in the history of mathematics. We will only be able to touch on the "big ideas."

A. The pivot point between the ancient Babylonians and Egyptians and modern mathematics occurs in the 17th century, when five different threads of mathematics come together—algebra, geometry, astronomy, mechanics, and the mathematics of motion.

B. We'll begin by looking at how different ancient civilizations—Mesopotamian, Egyptian, Greek, Indian, Chinese, and Islamic—developed these threads and laid the foundation on which western Europe would weave them together beginning around 1600.

C. We'll then spend five lectures on the 17th century, the point when these earlier mathematical ideas came to maturity and began to interact. One of the main figures in this interaction was Isaac Newton. His work *Mathematical Principles of Natural Philosophy* showed how all of the mathematics developed up to that point could be united.

D. Beyond the 17th century, we'll follow some of the most interesting developments and people in mathematics, with the intention of showing why it is so exciting to be a mathematician today.

Suggested Readings:

Courant, Robbins, and Stewart, *What Is Mathematics?*

Hamming, "The Unreasonable Effectiveness of Mathematics."

Wigner, "The Unreasonable Effectiveness of Mathematics in the Natural Sciences."

Questions to Consider:

1. Think of examples of quantities that are counted in discrete units (such as number of apples) and quantities that are measured in continuous units (such as quarts of applesauce). Speculate on how early people might have made the transition from numbers that count individual items to numbers that mark continuously varying and therefore necessarily arbitrary units.
2. Thinking back over your own mathematical education, identify as many points of contact as you can between algebra and geometry. When have you seen algebraic techniques used to illuminate geometric results; when have you seen geometric techniques used to illuminate algebraic results?

Lecture One—Transcript

What Is Mathematics?

Welcome to the history of mathematics, the "queen of the sciences." My name is David Bressoud. I'm a mathematician and a teacher, and I have a real passion for the history of mathematics because mathematics is really about ideas, and I think one of the most effective ways of understanding these ideas is to look at how they have developed. Where did they come from? What were people trying to do as they looked at these ideas? How did the ideas arise from many different civilizations and cultures? How did they interact? What were the difficulties that people encountered in developing them, and how do they play out in our world today?

I'm also interested in the history of mathematics because I'm interested in it as a teacher. As I try to explain these ideas to my students, I want to look back and be able to motivate the ideas by showing them where they came from, but also—and sometimes even more important—to look back and see where the difficulties have been. Where did people have trouble in coming up with a particular idea? What were the ideas that came naturally—that were easy to find? What were the ideas that were very difficult and took a long time to develop? Those places where the mathematicians and scientists had difficulty with the mathematics I can expect are going to be places where my own students are going to have difficulty. It greatly encourages them if I can say: "Look, this great mathematician, Augustin-Louis Cauchy, the greatest French mathematician of the 1820s, was struggling with this idea. If you're struggling with it too, that's just to be expected."

I also enjoy the history of mathematics just because it's full of great stories. We're going to be looking at a lot of interesting people. We're going to be studying people like Leonhard Euler, probably the greatest mathematician who ever lived, who would continue to be incredibly productive even after he was completely blind. We'll look at people who died extremely young: the young Evariste Galois, who would solve one of the most fundamental problems in mathematics at the age of 17 and die before his 21st birthday. We're going to look at the self-taught mathematician Srinivasa Ramanujan, and we're going to see how, despite the obstacles of growing up in southern India without access to much of the mathematics of western Europe at the time, he was able to get his hands on mathematics textbooks and learn from those and become one of the greatest mathematicians of the 20th century.

But before I get into the history of mathematics, I want to back up and look at the topic of this particular lecture, which is "What Is Mathematics?" To me, the essence of mathematics is the abstraction of pattern. What mathematicians do is to look at the world around us and see the patterns that are out there and then abstract those patterns in order to be able to work with them—and manipulate them, and use them, then, to tell us more about this world that we see around us. Some of the simple examples of this abstraction of pattern come from numbers. They come from geometry. Most people, when they think of mathematics, they think of numbers, and to some extent that is appropriate. I hope by the end of these lectures you'll have a much richer sense of the kinds of patterns that mathematicians work with.

But the beginning of understanding these patterns really is with number, so we look initially at number simply as a qualifier. We have 5 people. We have 5 loaves of bread. We are looking at 5 stars in the sky. From that, we abstract the number 5. That's very useful as we begin to combine objects. If I have 5 people in a room, and 3 more people come in, I now have 8 people. If I have 5 loaves of bread and add to those 3 more loaves of bread, I now have 8 loaves of bread. The fact that we've got, in essence, a 5 and a 3 that combine to become an 8 is something that goes beyond the people or the loaves of bread. It's a mathematical abstraction. Once we take those patterns and begin to abstract them, we can begin to play with them.

The same thing happens in geometry. The basic geometric objects are things like lines and circles, and they're really expressing spatial relationships. So we look at 5 stars in the sky, and we see that they seem to be in a certain pattern that we describe as a circle. We might look at a ring of stones on the ground, and we describe that as a circle. We might simply have 3 points that we're interested in. There's the tree over there, and the house that's here, and myself standing somewhere between them. I consider those 3 points, and I begin to think of them as a triangle. So we get this geometry that comes out of abstracting these patterns.

One of the interesting questions that comes up as you begin to think about these abstractions that mathematicians are working with is:

What is the reality behind them? Is 5 something that actually exists? Is a circle, a pure Platonic circle, something that really exists? Of course, I use the adjective *Platonic* because this gets back to Plato's view that, in fact, these ideal abstractions really do exist—that there is a reality beyond ourselves that we are tapping into as we make these abstractions. This gets to a more fundamental question that is often debated by mathematicians and philosophers of mathematics and philosophers of science—that is whether mathematics is discovered or whether mathematics is created. Are we simply revealing something that is out there as we put together mathematics, or are we creating something?

I actually come down between those people who believe that mathematics is discovered and mathematics is created. I am at heart a Platonist. I do believe that there is an ideal world out there that we are trying to understand, that we are trying to tap into. As we develop these abstractions of patterns that mathematicians work with, we're really trying to explain a deeper reality that is out there. It is also a very subtle reality, and most mathematicians, most scientists, realize that we can never really describe what's out there. What we are doing with mathematics is creating a language, a set of symbols, that enables us to work with some aspect of that deeper reality that is out there. But in most cases, we cannot claim that we really fully understand, that we really fully comprehend it. That's one of the things that makes mathematics so exciting, because as we abstract these patterns, we begin to try to look beyond what we've been working with.

What a mathematician does is to take an abstraction of a pattern that she or he sees over here and another pattern that's observed someplace else, and you look for points of similarity between these two patterns. As you overlay the patterns, as you begin to look for these points of similarity, you see that they suggest ways in which the pattern might extend—so you begin to explore that. Is this really going on? This is what makes mathematics so very exciting is this kind of exploration, knowing that there is a world out there that we understand very imperfectly, and we're exploring it in the mathematics—taking what we know, trying to find ways of combining what we know in order to be able to go further into the unknown and to understand that.

Now the patterns of mathematics come from many different sources, and I'll talk about many of these sources through this series of lectures. One of the most important sources comes to us out of commerce and civil administration. This certainly is where we get basic arithmetic. This is where we get our fundamental operations of addition, subtraction, multiplication, and division, and also simple fractions as well as the whole numbers, or integers. We can see ways in which fractions come up quite naturally. If I have 5 loaves of bread, and I want to share them among 3 people, then each person is going to get a loaf of bread and something left over. To be quite precise, each person, if they are shared equally, will get 1 and 2/3 loaves of bread—so fractions come in fairly naturally.

Commerce and civil administration also give rise to rates. What is the rate at which someone can accomplish a certain job? We want to know how long it will take to build a wall or dig a canal, and so you need to know the rate at which people are working. You want to know what the yield is on a given plot of land, and so you want to know how much barley is produced on each acre. Commerce and civil administration will also be the origins of much of the work in algebra. We're going to see how algebra comes out of these problems in arithmetic, but also how algebra comes out of problems in geometry. This will be one of the threads in some of the early lectures.

Another source of the patterns of mathematics comes from problems of navigation and problems of surveying. We're trying to do land measurements. This gives rise to geometry, which literally comes from the words *geo* ("earth") and *metri* or *metric* ("measurement")—so we are measuring the Earth. Geometry, then, arises from these problems of trying to determine distances, trying to determine areas, trying to determine volumes. As I said, we'll be using geometry to motivate much of what goes on in algebra, and we'll also be using geometry to motivate much of the number theory that I'll be talking about.

Number theory is really about understanding the structure of the integers. If I've got an integer like 20, how can I represent it? Can I write it as the sum of two squares? How can I decompose 20 into a product of primes? These kinds of questions are the questions that we're going to see addressed as we look at number theory.

A third source of the patterns of mathematics comes to us from astronomy or astrology. In the ancient world, there really is no clear distinction between astronomy and astrology. In fact, right up well into the 17th century, people would not make a distinction between these two. You studied the heavens in order to try to understand what was likely to happen on Earth. You really were doing astronomy, but it was done for astrological purposes. We're going to see how some of the greatest astronomers also were astrologers—in particular, people like Johannes Kepler at the very early years of the 1600s, whose work in both astronomy and astrology would lay the foundations for much of the work that would be done in developing the calculus.

This distinction between, or lack of distinction between, astronomers and astrologers also carries over into mathematicians. In fact, if you look at the ancient world, or even the medieval world, or even the Renaissance world, the words "mathematician," "astronomer," and "astrologer" were used interchangeably. One of the stories that I like very much is of the Emperor Tiberius—who, it is recorded, banished all "mathematicians" from the city of Rome. He was not banishing the accountants. He was banishing the astrologers. The astrologers were predicting his downfall, problems for his empire. He didn't like that fact, and so he was banning the astrologers. But to Tiberius—and for many hundreds of years after Tiberius—astronomy, astrology, and mathematics were all bound up in the same package. Largely, that's because some of the most important advances in mathematics came directly to us out of astronomy because we're trying to understand the patterns of the world around us, and most of the world around us is too messy to be able to pick out these patterns.

What we need to do is to take a look at a place where the patterns are much simpler, and astronomy would turn out to be one of those places. By looking at the heavens, we saw these patterns in a much purer form. But it's not just astronomy. We're going to continue that into a general viewpoint of the physical phenomena of the world and how we've pulled patterns from them. We will also be looking at optics, and eventually we're going to be looking at electricity and magnetism, and even the study of subatomic particles, and see how mathematics was influenced and how mathematics often was created from trying to explain these physical phenomena of the world around us.

Yet another source of mathematical patterns that will be very important is art and architecture. Especially as we get into the later lectures, we're going to see how the ideas of symmetry would come to play an absolutely fundamental role in the mathematics that would be developed in the 19th and the 20th centuries. That's one of the reasons why I'm standing in front of a picture of the Lion Court at Alhambra in Granada, because that particular palace is a wonderful example of symmetry in art, and it's actually through the symmetry at Alhambra that many mathematicians have come to a realization of the kinds of patterns that they then could apply to other problems—both in geometry and also in algebra.

I want to say a little bit more about number and distance. We've talked about number as abstraction from actual units, so we're counting the number of people or the number of loaves of bread. Sometime in far antiquity, somebody decided to take this idea of number and apply it to distance. You run into a problem there when you try to apply number to distance because the difficulty is there is no natural unit when you're looking at distances. If I want to know how far it is from here to there, well, today I might measure that off in feet, or meters, or yards, but those are simply human constructs. What are we going to use as the unit to measure off distance? Even once we decide what the unit is going to be, there's no reason why the unit we've chosen is really going to be appropriate for the distance that I'm looking at.

So if I've got my unit, say a stick, and I'm going to measure off the distance to some point, almost inevitably, as I get close to that point, I see that the distance that's left is less than the length of my stick. But I'm not able to take a whole number of sticks and measure that distance. One way of solving this problem of this little piece that's left over is to take my stick and mark it off into smaller intervals. I might mark it off into tenths, or I might follow the Babylonians and mark it off into 60 small pieces, or I might follow the Romans and mark it off into 12 small pieces. This is where we get our foot divided into 12 inches. I'm now going to take these smaller pieces and try to measure off the remaining distance.

But almost inevitably, the smaller pieces are not going to measure the remaining distance exactly, and so I'm going to need smaller units still. Take each of those inches, take each of those small pieces, and subdivide them. As I do that again, it won't exactly match.

Eventually, the scientists—or in the case of early mathematics, the bureaucrats—began to realize that, in fact, this was a process that would never end. Now the question is: Does it make sense to talk about the distance between here to there as a number if we can never express that number exactly—if the best we can do is get these approximations using smaller and smaller units? This is a question that the Greek mathematicians would really wrestle with, but we're also going to see it in the Babylonian mathematics of 4000 years ago.

Trying to apply number to time introduces other problems that are going to be very important to our story because, unlike distance where there is no natural unit, when I look at time we've got too many units. We've got the day, and we've got the lunar month, and we've got the solar year. Each of these units is incommensurable with the other. You cannot measure off a lunar month in a precise number of days. It's going to be a number of days plus a bit left over. You cannot measure off a solar year in an exact number of lunar months or an exact number of days. Whichever of these two units you use, there's going to be a little bit left over. As you subdivide these units, there's always going to be a little bit left over.

But even more importantly, what's happening is we've got different units traveling in different cycles, and this would present fascinating problems for mathematicians to wrestle with—how to figure out when all of these cycles would line up. A very important question in astronomy and astrology as people looked not just at the cycles of the day, the month, and the year, but also the period that it would take Mars to circle the Earth, or the time it would take Venus to circle the Earth, or the time it would take Saturn to circle the Earth—and then trying to find out when different planets would actually be in alignment in the future.

As we explore this question of the nature of mathematics, it's going to be very useful to refer to two papers that I'd like to talk about now—the first of which was written by the physicist Eugene Wigner, who taught at Princeton University. He published the paper in 1960; it's called "The Unreasonable Effectiveness of Mathematics in the Natural Sciences." One of the things that Wigner observed as a physicist was that physicists look at the world around them, and they try to understand the patterns that they see. Then they turn to mathematics, and they ask: Are there similar kinds of patterns that the mathematicians have been working with? Can I take this pattern in mathematics and apply it to the physical phenomenon?

Wigner remarked that, amazingly, most of the time when physicists see a pattern in the natural world, and they pick up a pattern from mathematics, that pattern in mathematics actually fits what's going on in the natural world with amazing accuracy—not just amazing accuracy; it also often has excess content. It is possible to take that mathematical model and see beyond what we were trying to explain. So not only does the mathematical model enable us to do the predictions of what we already saw, it suggests a deeper reality that is out there. I will return to this theme periodically as I talk about the work of Newton—and also especially in the second-to-last lecture, when I'm going to show how mathematics really does represent the ultimate physical reality, showing us things that we never would have expected without the mathematical models.

Wigner asks the question: Why is mathematics so good at this? He defines mathematics as "the science of skillful operations with concepts and rules invented just for this purpose." That suggests that there's an arbitrary nature to mathematics that I don't quite buy. I do not believe that mathematicians just sit at their desks and try to think up interesting patterns that they can play around with. As I said, mathematicians really get their ideas for the patterns that they will study by looking at the world around them. So in some respects, we should not be surprised that if we look at the world around us, and we extract a pattern, and then we try to apply this pattern to the world around us, it often will fit.

But what is surprising is that the patterns that are pulled out of one area—so for example, we're going to see how astronomical phenomena gave rise to the trigonometric functions. We can then take those trigonometric functions and apply them to something that seems to be totally unrelated to the heavens—namely, a vibrating string—and see that exactly these trigonometric functions that came out of our study of astronomy are going to be applicable to problems in mechanics, problems of vibrating strings, and actually eventually applicable to problems like heat flow and electricity and magnetism.

There was a second article that I want to talk about. This is really a response to Wigner's article. It was written by Richard Hamming in 1980, and it's called simply "The Unreasonable Effectiveness of

Mathematics." Richard Hamming was a very prominent applied mathematician. He worked at Bell Labs, and he did a lot of work in signal processing. He was also interested in this question of why mathematics seems to be so useful, so effective. He did raise that point that I did—that after all, mathematics is extracted from the world around us, so we shouldn't be too surprised that it's applicable when we try to apply it back to the world. He was actually less surprised, though, that it could be applied to other areas than I am.

I want to raise Hamming's article because he does a nice job of clarifying what mathematics is. He describes the four faces of mathematics, and the first two I have already talked about. The first face of mathematics is number. The second face of mathematics is geometry. The third face of mathematics is close reasoning. This is extremely important because, as I said, as we're looking at mathematics, we're taking these patterns—we're overlaying patterns—and very often we're extracting patterns from the overlay of patterns. As we build these abstractions of the patterns that we saw, they create new patterns, and we abstract from them. Mathematicians are often working with abstractions of abstractions of abstractions of abstractions. To keep track of what's going on and to really have a reliance on the results that you get—to have some confidence that you really have found something that really is there, that really will be useful—you need tools in order to keep track of these abstractions of abstractions and how they work together, and what are the legitimate conclusions that can be drawn, and what are not the legitimate conclusions.

One of the ways of doing this is through formal logic, the logic that was developed in Greece, which comes to us from the Greek mathematical tradition. But I like the fact that Hamming uses the words "close reasoning" rather than "logic," because I think the picture is much broader than just the kind of formal logic that we know from Euclid's *Elements*, the Euclidean geometry. As we look at many different cultures, many civilizations, and the way that they have developed mathematics, we're going to see that they came up with different ways of maintaining this close reasoning. It did not have to be the kind of formal logic developed in Greece.

The fourth of the faces that Hamming talks about is artistic taste, and I really like this because there is something about mathematics that is very aesthetically appealing. As we're looking for these patterns, there almost inevitably has to be an aesthetic appeal if we're on the right track. When we see the kinds of symmetries that we want, when we see these kinds of patterns coming together and just clicking perfectly—perhaps in an unexpected way, but in a very satisfying way—that's what tells us that we are working with real mathematics and with important mathematics. It's very difficult to define this "aha" moment when things come together, but it's incredibly exciting when it does. There is this real feeling of the need for artistic taste and the artistic appreciation that comes from it. I believe that mathematics is far less arbitrary than Wigner talked about it. I believe that it's much more surprising than Hamming claimed it was, and I hope to be able to communicate that to you in this series of lectures.

My lectures are going to cover 4000 years of the history of mathematics, and there's no way I can do this in any kind of detail. I want to bring you right up to the 21st century. I want to share with you some of the exciting things that mathematicians are doing today—so I'm going to have to treat much of this history of mathematics in a very superficial way. But I hope to be able to communicate some of the really big ideas. In looking at how to structure these 4000 years, going all the way back to the ancient Babylonians and Egyptians, I decided that the real pivot point occurs in the 17th century, and that's when we get five different threads coming together. The threads are algebra, geometry, astronomy, and mechanics—the way that physical systems interact with each other—and then the last of these is going to be the mathematics of motion.

We're going to see how different ancient civilizations developed these threads until—by around the year 1600—they come together. We're going to see these threads coming out of the ancient Mesopotamians, the Egyptians, and the Greeks. We're going to go to South Asia and see how the Indian mathematicians of the 1st millennium A.D. developed the ideas. We're going to turn to China and see how these ideas were further developed there, how the Islamic world took all of the different threads of mathematics that had been developed up until that time and really pushed them much, much further and laid the foundation on which western Europe would be able to pick up these threads and continue to develop them until, by 1600, they were ready to be woven together.

I'm actually going to spend five entire lectures just on the 17th century because this is such an exciting time, as these mathematical ideas came to maturity and people began to see how they could interact, how you could weave them together. The great master of weaving these threads together was Isaac Newton. He saw how to combine them in new and very important ways. Newton's *Mathematical Principles of Natural Philosophy*—not truly a work of mathematics; it's really a work of physics, or of science—is explaining celestial mechanics. In it, he shows how all of the mathematics that had been developed up to that point could be brought together and united into very exciting new mathematical ideas.

Beyond the 17th century, I will follow the development of mathematics—treating it very superficially now, just touching on a few of the points that I find most interesting, some of the people that I enjoy talking about the most. The mathematics after the 17th century is going to be extremely personal. I'm going to share with you my own enthusiasms, my own excitements, and hopefully bring you up to the point where you have some idea of why it is so exciting to be a mathematician today.

Lecture Two

Babylonian and Egyptian Mathematics

Scope:

This lecture begins with an overview of the sources for the earliest recorded mathematics, from the Old Babylonian period (2000–1600 B.C.) in Mesopotamia and the Early Middle Kingdom in Egypt. Both civilizations had highly developed though different systems for recording and calculating with numbers, including fractions. Both civilizations knew how to find areas and volumes, including the area of a circle, and both used the Pythagorean theorem. The Babylonians demonstrated knowledge of how to generate arbitrarily many Pythagorean triples and had sophisticated techniques for calculating square roots and solving area problems that are equivalent to the solutions of quadratic equations.

Outline

I. In this lecture we will go back almost 4000 years, to the earliest recorded mathematics, which can be found in ancient Mesopotamia and Egypt.

- **A.** Sophisticated mathematical texts were developed in both these civilizations at about the same time; these texts show us the origins of algebra and of the idea of approximating values.
- **B.** In Mesopotamia, we will look at what is known as the Old Babylonian period, from roughly 2000 to 1600 B.C. In Egypt, we will look at the Early Middle Kingdom, dated to approximately the same period.
- **C.** During this time, both of these civilizations became increasingly sophisticated. They had highly developed systems of commerce and were engaged in large-scale construction. They needed a large corps of people trained in basic mathematics to manage taxation and oversee building projects.

II. The records of mathematics we have from these ancient civilizations are the textbooks used to train such bureaucrats. These are primarily sets of problems with solutions.

A. From Egypt, we have only two problem books.

1. One of them, the Rhind Papyrus, was probably copied around the 17th century B.C. by a scribe named Ahmes. That scribe developed the text from a manuscript believed to have been written 200 years earlier. The papyrus roll is about 13 inches wide and 18 feet long.
2. The other important source for Egyptian mathematics from this period is the Moscow Mathematical Papyrus, probably written in the 19th or 18th century B.C. Unfortunately, only a small piece of this papyrus still exists today.

B. These two problem books show us how the ancient Egyptians represented number.

1. Our modern representation of number is quite sophisticated. We use 10 digits, the digits 0 through 9; if we want to represent a number that is greater than 9, we reuse these digits, but we put them in places that carry extra value.
   - **a.** For example, the two digits 1 and 3 can be combined to write 13 or 31, but the 3 may represent three 1s or three 10s, depending on where it is placed in the written representation.
   - **b.** In our modern system, the original 10 digits can be put in different positions to indicate the number of 10s, 100s, 1000s, and so on.
2. The system used in ancient Egypt was much simpler. This system represents numbers with strokes: 1 is a single stroke, 2 is 2 strokes, and so on.
   - **a.** We still see echoes of this in our modern representations of numbers. The written numeral 2, for example, is based on 2 horizontal strokes connected by a curve.
   - **b.** The stroke system becomes cumbersome in representing larger numbers. Thus, the Egyptians devised a system that replaced a group of 10 strokes with a single symbol called a *yoke*. A group of 10 yoke symbols was replaced with a different symbol, a *rope*.

c. A group of 10 rope symbols would be replaced with a lotus symbol to represent 1000, and other symbols were used for 10,000 and 100,000.

C. The Egyptians were also able to work with simple fractions by thinking of them as the reciprocals of integers.

1. The fraction $\frac{1}{2}$, for example, would be represented by 2 strokes with an oval drawn over the top of them.
2. A fraction like $\frac{3}{4}$ would be represented as $\frac{1}{2}$ plus $\frac{1}{4}$. This system was sufficient for the uses of fractions in ancient Egypt.
3. One of the problems from the Rhind Papyrus asks the following: We have a scoop and a certain volume of wheat called a *hekat*. If it takes $3\frac{1}{3}$ scoops to equal 1 *hekat*, what is the size of 1 scoop?

III. The ancient Egyptians were proficient in their work with geometry. They were able to find formulas for areas and volumes, and they knew the Pythagorean theorem.

A. The Pythagorean theorem is clearly much older than Pythagoras, who lived in the 6th century B.C. Even by 2000 B.C., people in Mesopotamia and Egypt were conversant with the Pythagorean theorem.

1. The Pythagorean theorem relates to areas. To understand the theorem, imagine starting with a right-angle triangle and constructing a square on each of the three sides. The Pythagorean theorem states that the areas of the two smaller squares, when added together, are precisely the area of the large square.
2. Today, we state the Pythagorean theorem as: The square of the length of one of the legs of a right-angle triangle plus the square of the length of another leg of the triangle is equal to the square of the length of the hypotenuse ($a^2 + b^2 = c^2$).
3. The literal meaning of the word *square* in this statement helps us understand the relationship of the theorem to areas.

B. The Egyptians were also able to find the area of a circle.

1. The problem of finding areas of regions that are not square or rectangular was simplified to the problem of finding the side of a square whose area was equal to the given area. This approach, called *squaring a circle*, was used by Egyptian, Babylonian, Greek, medieval, and Renaissance mathematicians.
2. With this approach, we start with a circle of a certain diameter and then try to find the side length of a square that has the same area as that circle.
3. The Egyptians found that constructing a square that is $\frac{8}{9}$ the diameter of the original circle would give an extremely close approximation to the actual area of the circle.

IV. For the ancient Babylonian Kingdom in Mesopotamia, we have thousands of records of mathematics, but these are usually fairly small clay tablets, containing perhaps a single problem with its solution.

A. One of these tablets, known as Plimpton 322, is about 3.5 inches high and 5 inches long and is believed to have been created about 1800 B.C. This tablet lists cases of the Pythagorean theorem in which all three sides of the triangles are integers. Such cases are called *Pythagorean triples*.

1. The integers (3, 4, 5) make up one example of a Pythagorean triple ($3^2+4^2=5^2$). Other examples include (5, 12, 13) and (8, 15, 17).
2. Amazingly, among the triples listed on Plimpton 322 are (119, 120, 169), (3367, 3456, 4825), and (4601, 4800, 6649).
3. Unfortunately, Plimpton 322 consists only of this table of values. We have no idea of the procedure used to create it.

B. For small numbers, the Babylonians represented numbers in the same way that the Egyptians had.

1. A vertical wedge symbol was used to represent numbers up to 10, and a horizontal wedge represented 10. Rather than continuing up to 100, however, the Babylonians stopped at 60.

2. When they reached 60, instead of using a new symbol, the Babylonians reused the symbol for 1, in the same way that we reuse the symbol for 3 in the 10s place to mean 30.
3. Instead of our base-10 system, the Babylonians used a base-60 system. A symbol for 1 in the 60s place meant 60. A mathematician could then build up 60s to sixty 60s, then start over again with the symbol for 1 in the place of sixty 60s.
4. Thus, as we have a 1s place, a 10s place, and a 100s place, the Babylonians had a 1s place, a 60s place, and a 3600s place.
5. Look at the number 4601 as an example. There is one 3600 in 4601. Taking out 3600 leaves 1001. There are sixteen 60s (960) in 1001. Taking out 960 leaves 41. Thus, the number 4601 would be represented by one 3600, sixteen 60s, and forty-one 1s.

C. Another Babylonian tablet, known as YBC 7289, dates from between 1800 and 1600 B.C. and shows a representation of $\sqrt{2}$.
   1. The problem this tablet explores is as follows: Consider a square and the diagonal of the square. If we take the length of the square as our unit, how many times can we measure off the length of that square against the diagonal?
   2. If we try to perform the measurement, we realize that the length of the square will go into the diagonal once but not twice. How big is the piece of the diagonal that remains?
   3. By breaking down the original unit, the length of the square, into groups of 60 progressively smaller subunits, the Babylonians were ultimately able to arrive at a very close approximation to $\sqrt{2}$. In our modern decimal representation, their solution is approximately 1.414213.

D. We can also get an idea of how the Babylonians viewed algebra problems from ancient clay tablets. In fact, the Babylonians did not think about algebra in the same way that we do today, but they did share common approaches with Greek, Indian, and Chinese mathematicians.

1. One of the problems from a Babylonian tablet is as follows: We have a square with a side of unknown length. We want to add to that the length of the square. The sum of these values is $\frac{3}{4}$; what is the length of the original square?
2. Today, we would find the solution with an algebraic equation: $x^2 + x = \frac{3}{4}$.
3. The Babylonians, however, thought about this problem geometrically, creating and manipulating rectangles to arrive at a method of proof that is exactly equivalent to the way we would solve this problem algebraically today. The technique for solving quadratic equations is called *completing the square*.
4. The Babylonian approach shows us the power of studying the history of mathematics. Learning to solve quadratic equations is easier if students are able to think about completing the square in terms of geometry and then encode that idea in algebraic notation.

V. After about 1600 B.C., the ancient sources for mathematics are not nearly as rich as they are for earlier periods. To some extent, mathematics seems to have gone underground at this point, taken over by the priestly castes for use in astronomical observations.

A. Nonetheless, we believe that the mathematical tradition became deeper and richer in both Mesopotamia and Egypt, largely because it would resurface in the Greek world, starting in about 600 B.C.

B. In the next lecture, we'll jump ahead to 600 B.C. to see how the Greeks used the ideas they inherited from Mesopotamia and Egypt.

Suggested Readings:

Buck, "Sherlock Holmes in Babylon."

Gillings, *Mathematics in the Time of the Pharaohs*.

Høyrup, *Lengths, Widths, Surfaces*.

Katz, *A History of Mathematics*, chap.1.

Van der Waerden, *Science Awakening I*, chaps. 1–3.

Questions to Consider:

1. The bureaucrats of the early 2nd millennium B.C. in Mesopotamia and Egypt needed to be skilled at engineering, law, and business. What are some of the mathematical skills they needed to perform their tasks?
2. We have no evidence of proofs in the modern mathematical sense in early Babylonian or Egyptian mathematics. What are the circumstances that would create a need for proofs?

Lecture Two—Transcript

Babylonian and Egyptian Mathematics

In this lecture, we are going to go back to the earliest recorded mathematics, going back almost 4000 years—to both ancient Mesopotamia and ancient Egypt. In both of these civilizations, at about the same time, we get the development of quite sophisticated mathematical texts, and we are going to be looking at these texts for the origins of the ideas of algebra, and also the origins of the idea of approximating values.

For Mesopotamia, we are looking at what is known as the Old Babylonian period, roughly about 2000 B.C. to around 1600 B.C. In Egypt, we are looking at the Early Middle Kingdom, also over approximately the same period. What was happening in both of these civilizations—both Mesopotamia, in what today is modern Iraq, and also in Egypt along the Nile River Valley—is that both of these were getting very sophisticated civilizations that had highly developed systems of commerce, and they were doing a lot of construction. They needed to train a large corps of civil servants, of bureaucrats, of people who could oversee taxation, who could oversee the building projects that were coming about at that time—and these were people who needed to be trained in basic mathematical techniques.

The mathematics that we have from this period are the textbooks that would have been used in order to train these people. It is not mathematics the way we might like to see it—which is where did they get these ideas, and how did they develop these ideas—but what we get are problem books, sets of problems, together with worked-out solutions. These are problems that students would have worked on and tried to emulate.

The problem books that we have that come to us from Egypt are really quite limited. There really are only two of these books, but they are both quite complete—and especially one of them. We've got the Rhind Papyrus that was probably written around the 17th century B.C.; the name of the scribe who wrote it was Ahmes, and we do know that in fact this was developed from, or copied from, a slightly earlier manuscript believed to have been written about 200 years earlier. It is quite complete. The actual papyrus roll that exists is about 13 inches high and about 18 feet long.

There is one other important source for Egyptian mathematics from this period, and that is the Moscow Mathematical Papyrus, which was probably written in the 19th to the 18th century B.C., that exists today in the Pushkin State Museum. But unfortunately we only have a small piece of that papyrus; only the top edge still exists. It is about the top 3 inches of it, but the whole papyrus is about 15 feet long.

We have basically got these two problem books that lay out the kinds of problems that civil servants in ancient Egypt were expected to be able to do. What is very interesting is the way that the ancient Egyptians represented number. The way we represent number today is really quite sophisticated. What we do with number is we have 10 symbols, the 10 digits, 0 through 9, and if we want to represent a number that is larger than 9, we start reusing these digits, but we put them in places that carry this extra value. For example, the two digits 1 and 3 can be combined to 13, or we can also combine the two digits 1 and 3 into 31. The 3 has a very different meaning depending on whether it is 13 or whether it is 31. In the latter case, 31, that digit 3 represents 30; it is three 10s. And so we take these 10 digits, and we put them in different positions in order to indicate how many 10s we have, or how many 100s we have, or how many 1000s we have.

The system that was used in ancient Egypt is a much simpler system, and it is based on the idea, first of all, starting out by representing a number by strokes, which is certainly the very oldest way of representing a number: 1 is a single stroke, and 2 would be 2 strokes, and 3 would be 3 strokes, and 4 would be 4 strokes. We still see some echoes of this in our modern way of representing numbers, because the digit 2, for example, is really based on 2 horizontal strokes with a curve connecting them. The number 3 is really based on 3 horizontal strokes with curves connecting them.

It becomes very cumbersome if you want to represent every number by strokes, and so what the Egyptians devised was a system by which once they got up to 10 of these strokes, they would replace 10 strokes with a single symbol—what we call a *yoke*. A yoke was a single symbol that would represent 10. Once you got up to 10 of these symbols for 10, then you would replace them by a single symbol, referred to as a *rope*, and so it is simply a spiraling symbol that meant you have got 100. So if you wanted to represent 317, for example, you would have 3 of the rope symbols, 1 of the yoke symbols, and 7 of the stroke symbols. If you wanted 10 of the ropes, then that was represented by a lotus symbol that represented 1000, and so on, and this was continued up into 10,000 and 100,000.

In addition to that, the Egyptians were able to work with simple fractions. What they did was to think of the reciprocals of the integers, and so if you wanted to represent 1/2, you would simply take 2 strokes and draw an oval over the top of them, and so that represented the reciprocal of 2, or 1/2; 1/3 would be 3 strokes with an oval over the top of it. You have got a problem here in that how do you represent a number like 3/4? What the Egyptians would do is represent 3/4 as 1/2, so 2 with an oval over it, plus 1/4, so that would be 4 with another oval over that. That really was sufficient for the kinds of uses of fractions that you get in ancient Egypt.

The kinds of problems that they would work with, one of the examples from the Rhind Papyrus is that you have got a scoop, and you've got a certain volume of wheat, a *hekat*, and the problem that was posed was: It takes 3 and 1/3 scoops in order to equal 1 *hekat*, so what is the size of a single scoop? This is one of the problems that these people who were training to be civil servants would practice.

The ancient Egyptians were quite good in their work with geometry. They were able to find a lot of formulas for areas, for volumes, and they knew the Pythagorean theorem. In the next lecture, I am going to talk about Pythagoras. He lived in the 6th century B.C., so the Pythagorean theorem is definitely much, much older than Pythagoras himself—in fact it is clear that even by the year 2000 B.C. in Mesopotamia and in Egypt, people were very conversant with the Pythagorean theorem. They were using it in many different contexts, and so it must be much, much older than 2000 B.C.

The Pythagorean theorem—the way to think about it is the way that really has been thought about it throughout most of the history of mathematics. The Pythagorean theorem is really a theorem about areas. You take a right-angle triangle, and you construct a square on each of the three sides. You take the bottom side and construct a square on that, and you take one of the other sides and construct a square on that, and you take the third side and you construct a square on that, and what the Pythagorean theorem says is that the areas of the two smaller squares, when added together, give you precisely the area of the large square. Today we know the Pythagorean theorem as: The square of the length of one of the legs plus the square of the length of another leg is equal to the square of the length of the hypotenuse ($a^2 + b^2 = c^2$). But what is really going on there, when we use the word *square* it should be taken literally, and that is the way people have thought about this theorem, right up until very recent centuries, as a theorem about areas.

Another of the problems that the Egyptians solved was that of finding the area of a circle. The problem of finding areas of things that are not square regions or rectangular regions was always simplified into the problem of finding the side of a square whose area is equal to the given area. So this is a problem that would be stated not just by the Egyptians and the Babylonians but also by Greeks and also by medieval and Renaissance mathematicians, the problem of *squaring a circle.*

You have got a circle of a certain diameter, and you want to find the side of a square that has the same area as that circle. One of the amazing results that the ancient Egyptians actually were able to come up with is, the way to find the side of that square is to take the diameter of the circle and take 8/9 of that diameter. So you construct a square whose side is 8/9 of the original diameter, and that actually gives you an extremely close approximation to the actual area of the circle—an approximation that is off by less than 1 part in 100, certainly accurate enough for the kind of work that these Egyptian scribes needed to be able to do.

The situation in the ancient Babylonian Kingdom and Mesopotamia was quite different from what was going on in Egypt at that time. As I said, from Egypt we have basically got two sources. From Mesopotamia we have got literally thousands of sources, but they are not complete texts—they are tiny fragments of texts because in Mesopotamia, rather than writing on papyrus scrolls, what the scribes did was to record information on clay tablets. They would take a clay tablet, and they would take a reed stylus, and they would stamp marks in the clay tablet. The clay tablets generally were fairly small, and so very often a clay tablet would just contain a single problem with its solution, or perhaps it would be a table of values, but you would not be able to get a complete math text, but we have thousands of these clay tablets that have survived, and so we have got a very good idea of what was happening in Babylonian mathematics at this time.

I want to focus on two of those tablets because they give a good indication of the kinds of mathematics that the Babylonians were able to do. The first of these is known as Plimpton 322, and today it resides at Columbia University, and it is about 3.5 inches high, about 5 inches long, and it is believed to have been created about 1800 B.C. This is a tablet that deals with the Pythagorean theorem, and in particular it is looking at cases of the Pythagorean theorem where you can take the three sides of the triangle to all be integers.

These are called *Pythagorean triples*. This is three integers (such as 3, 4, 5) where 3^2 (which is 9) + 4^2 (which is 16), 9 and 16 is 25, and 25 is a perfect square. Another example of a Pythagorean triple is (5, 12, and 13). Another example is (8, 15, and 17). Sometime back in ancient prehistory people began, after studying the Pythagorean theorem and realizing what was going on there, looking for triples of integers that had this property.

What is amazing about Plimpton 322 is that it does not just find triples like (3, 4, 5) and (5, 12, 13), but among the triples that are contained on Plimpton 322 are (119, 120, and 169). If you take $119^2 + 120^2$, that is equal to 169^2. It is very unlikely that anybody could find this just by chance, and if you have got any question about what is going on here, some of the other triples that occur on Plimpton 322 are 3367, 3456, and 4825. Another triple is 4601, and if I square that, 4800, square that and add it to the first square, and I am going to get a perfect square, exactly 6649^2.

There must be something going on behind this. Unfortunately, for Plimpton 322 all we have is this table of values. We have no idea how it was created. There is some fascinating mathematical detective work trying to figure out what else we can pick up from what other things we see on Babylonian mathematics to figure out how they might have come up with these triples, but clearly they must have had a general procedure for coming up with them.

This is a good opportunity to also talk about the way in which the Babylonians represented numbers. They started out in the same way that the Egyptians had, so they had a particular symbol, a vertical wedge, that represented 1, and 2 of these would represent 2, and 3 of them would represent 3, until they got up to 10, and then they had a new symbol for 10, and that was simply a wedge put on its side—a sideways wedge. A horizontal wedge represented 10, and then you built up from there. But rather than going all the way to 100, the Babylonians stopped at 60. We are not exactly sure why 60, but my guess is it has something to do with the fact both that there are approximately 360 days to a year, and also the fact that 60 is a number that is divisible by many nice small integers.

They worked with 60, and then instead of coming up with a new symbol for 60, what they did was to reuse the symbol for 1, in the same way that when I am writing 31, I reuse that 3 symbol, but in the 10s place. The Babylonians, instead of having a base-10 system—based on powers of 10 as our modern numbers are—they had a base-60 system, based on powers of 60. So if they would put a 1 in the 60s place, that would then indicate that they had 60. They could build up 60s, and once they got up to sixty 60s, they would start over again, again with their symbol for 1 that would go into the place of sixty 60s, and sixty 60s is 3600. They would have a units place, and they would have a 60s place, and they would have a 3600s place, and they could continue that.

For example, for the number 4601, there is one 3600 inside of that. You take out 3600, and you are left with 1001. There are sixteen 60s in that. You take out the sixteen 60s, which is 960, and you are left with 41. So they would represent this number—4601—by 3600, sixteen 60s, and forty-one 1s.

The next piece of Babylonian pottery, the next mathematics that I want to talk about, is called YBC 7289. It is a rather unromantic name, but that happens to be its designation from the archeological dig that it was discovered in. It exists today at Yale University. It probably dates from between 1800 and 1600 B.C., and what is going on on this particular piece of mathematics is that the Babylonians are representing $\sqrt{2}$.

We consider a square, and we look at the diagonal of the square, and the problem that they are really wrestling with is if I take the length of that square as my unit, how many times can I measure off the length of that square against the diagonal? You try to measure off the length of the square against the diagonal, and you realize that it goes in once and it is not going to go in twice—so there is a piece left over. How big is that piece? What they did then was to take their unit and break it down into 60 subunits, so sixtieths of their basic unit. This idea of taking their units and breaking it down into sixtieths, incidentally, is something that we still have with us today. The fact that we break an hour down into 60 minutes and a minute down into 60 seconds comes directly to us out of the Babylonian way of dealing with fractions.

The Babylonians would deal with sixtieths, and if you take the piece on the diagonal that is left over, it is about 24/60, but not exactly. You can get 24/60, and then there is a little piece left over. On this clay tablet, what the Babylonians did was take their sixtieths of a unit and break it down into sixtieths of sixtieths. They saw that there were about 51 of these sixtieths of sixtieths, but there was a little piece left over, and they then analyzed the little piece that was left over, and they found that there were about 10/60 of sixtieths of sixtieths.

You are not going to be able to do that by actually measuring, and in fact you really can't get much beyond just sixtieths of a unit, so there had to be some very sophisticated algebra that was sitting behind what the Babylonians were able to do in order to get this approximation to $\sqrt{2}$ so very accurately. The approximation that they came up with was that $\sqrt{2}$ was approximately $1 + 24/60 + 51/(60 \times 60) + 10/(60 \times 60 \times 60)$, which in our modern decimal representation is approximately 1.414213. This is incredible accuracy. The Babylonians were able to get $\sqrt{2}$ down to an accuracy of less than 1 part in 1 million.

Not only did they deal with this question of how to find approximations, there also is a lot of algebra that is contained in the Old Babylonian clay tablets. It is not algebra the way we think about algebra today, and in fact one of the big themes that we will be following through this series of lectures is how algebra actually did develop, but it is very instructive to look at the way that the Babylonians viewed algebraic problems—because that is the way the Greeks would look at it, it is the way the Indians would look at it, and it is the way the Chinese would look at it. It really is not up until about the year A.D. 1000 that we begin to get algebraic notation and algebraic ideas that really are connected to our modern way of thinking about algebra. The kinds of problems that the Babylonians looked at were very algebraic, and one of the problems that is contained on one of the Babylonian tablets is the following.

You have got a square of unknown side, and we want to add to that the length of that square, and the value is 3/4—what is the length of the square? Today the way we would set that up is with an algebraic equation. We would say $x^2 + x = 3/4$, and we would solve it algebraically, but the algebraic notation did not exist at this time. What the Babylonians did was to think about this problem geometrically. You've got a square, and you've got a certain length that is equal to that square—let's create a second area to add to the square. We will create a rectangle whose height is equal to the length of the square and whose width is 1 unit, so that the area of this rectangle is exactly equal to the length of the original square.

In our modern algebraic notation, if we think of the side of the square as being x, x by x, then our rectangle is x high and it is 1 unit wide. The method of proof is exactly equivalent to the way we would solve this problem algebraically today. What the scribe is told to do is take that rectangle and break it in half—so you get two rectangles, each of height this unknown quantity, and each of width exactly 1/2. We take one of those rectangles, and we slide it underneath our square—so now I've got my square, I've got a rectangle over here of width 1/2 and height equal to the height of the square. I've got another rectangle underneath the square whose width is equal to the length of the square and whose height is equal to 1/2, and now we complete the square. *Completing the square*, for the ancient Babylonians and for the Greeks, again, right up until relatively recently, is filling in the missing area on an actual geometric square. We want to fill in that little missing piece that is down in the corner, and we know exactly how big that little piece is: It is exactly 1/2 a unit high and 1/2 a unit wide, and so its area is equal to $1/2 \times 1/2$, which is equal to 1/4.

That tells us that if I add an area equal to 1/4 to my existing area, I am going to get an exact square. I knew that the area originally was 3/4, and if I add an area of 1/4, that now gives me an area of 1, and so my entire square has area 1. It is easy to determine what the side of that square is—it must be a square of side 1, but I know that the side is also equal to the unknown side of the original square plus 1/2, and that tells me that the original square must have been 1/2 a unit by 1/2 a unit.

That is exactly what we do algebraically today, and this is one of the reasons in the first lecture why I talked about the power of looking at the history of mathematics, because teaching students how to solve a quadratic equation by completing the square suddenly makes a lot more sense if you are actually able to think about it geometrically and not think about completing the square as some abstract operation but actually as taking a geometric square and filling in the missing piece. Once you understand what is going on geometrically here, you can then take that idea and see how to encode it in the algebraic notation, but it would take 3000 years before these problems of solving quadratic equations got encoded in the kind of algebraic notation that we are familiar with today.

I have been looking at Babylonian mathematics and Egyptian mathematics basically in the period 2000 to 1600 B.C. I have been looking at the kinds of text that were being used to train the scribes, to train the civil administrators, and you do not get a lot of mathematics; you get some mathematics after 1600 B.C., but the sources are not nearly as rich as they had been before, and to some extent much of the mathematics goes underground at this point. Much of the mathematics in Babylon and Egypt is taken over by the priestly castes, who are using mathematics for astronomical observations.

I am going to wait until Lecture Five before I really begin to look at astronomy and its effect on mathematics in any detail, but one of the things that is important to remember is that the mathematical tradition continues both in Mesopotamia and Egypt. We have every reason to believe that it becomes much deeper and richer—largely because we know that it has direct influence on what would come into being in the Greek world, starting in about the year 600 B.C.—but it does become secret knowledge.

This ability to understand what was going on in the heavens was understood as a way of predicting what was likely to happen on Earth. After all, the position of the Sun in the heavens certainly has great influence on what is going on on the Earth, and so very likely the position of the planets, the position of the Moon, should also have an influence on what is going on, and this then would become secret knowledge. In Mesopotamia, these mathematicians, astronomers, and astrologers would come to be called magi, and the Magi who visit the infant Jesus in the Christmas story are mathematicians—Persian mathematicians from Mesopotamia who have seen a star and have realized the astrological implications of the star that they saw and follow it. We get mathematicians appearing in the New Testament.

In the next lecture, we are going to jump ahead 1200 years to about 600 B.C., and we are going to see how the Greek mathematicians of that time picked up the ideas that they inherited from Mesopotamia and from Egypt and then develop that mathematics in very new and very important directions.

Lecture Three

Greek Mathematics—Thales to Euclid

Scope:

This lecture surveys more than 300 years of Greek mathematics, beginning with Thales of Miletus and continuing through Pythagoras to Zeno, Aristotle, Theaetetus, and Eudoxus. We will explore some of the mathematics developed by these thinkers, such as Thales's calculations of distances, the Pythagoreans' formulation of the idea of irrational numbers, Zeno's paradoxes, Aristotle's division of mathematics into *arithmos* and *logos*, Theaetetus's work with the Euclidean algorithm, and Eudoxus's procedure for determining that two irrational numbers are the same. We then turn to Euclid's *Elements*, considered to be the most important book of mathematics ever written. The *Elements* encapsulates earlier Greek mathematics, including solid geometry, properties of ratio and proportion, number theory, and prime numbers.

Outline

I. The Hellenistic world would dominate mathematical development from about 600 B.C. to A.D. 400. This lecture focuses on the period from 600 to 300 B.C., about the time when Euclid of Alexandria wrote his great books on mathematics.

- **A.** Euclid is best known for the *Elements*, which is often considered to be a work of geometry but actually covers much more mathematics.
- **B.** Scholars debate the extent to which the theorems recorded by Euclid were original to him.
- **C.** Euclid's *Elements*, however, did summarize everything that was then known about mathematics and set the foundation for everything that would come afterward.

II. One of the earliest Greek mathematicians was Thales of Miletus (c. 624–c. 545 B.C.), who was interested in calculating distances between objects.

- **A.** We know that Thales studied with Egyptian mathematicians and astronomers. He is said to have predicted an eclipse that occurred in the year 585 B.C. and calculated the height of the pyramids.

B. According to Plato, Thales also discovered a geometric method for determining the distance to a ship using two different points of observation on the shoreline. His fascination with the idea of distance set an important example for Greek mathematics.

1. Although we see some problems related to commerce and astronomy in Greek mathematics, the geometric interests of Thales would come to dominate.
2. This focus on geometry is evident in the dependence of the Greeks on the idea of ratio and proportion; often, instead of working with pure numbers, Greek mathematicians chose to work with ratios.

III. Ratios play a major role in the work of Pythagoras of Samos (c. 580–c. 500 B.C.).

A. Scholars believe that Pythagoras probably met Thales and may have even studied with him. He studied mathematics and astronomy in both Egypt and Babylon and was indoctrinated into many of the mystic practices of the Egyptian astrologer-priests.

B. After leaving Babylon, Pythagoras went to Italy, where he founded his own school based on many of the cult practices of Egyptian and Babylonian mathematicians.

1. The Pythagoreans believed that mathematics lay at the core of all nature. According to Plato, Pythagoras said, "At its deepest level, reality is mathematical in nature."
2. The Pythagoreans were particularly interested in expanding their knowledge of mathematics built around ratio and proportion. Certain developments in music and musical scales, as well as in architecture, are attributed to this group.
3. Of course, the Pythagorean theorem is often attributed to this group—but as mentioned in the last lecture, it was known much earlier. It appears, however, that the Pythagoreans were the first to develop a proof of the theorem.

C. The Pythagoreans are also credited with formulating the idea of irrational numbers—that is, numbers that cannot be represented using fractions or a ratio of two integers.
   1. The first example of an irrational number we know of is $\sqrt{2}$, which is approximately equal to $\frac{7}{5}$. No fraction is exactly equal to $\sqrt{2}$.
   2. The proof of the fact that $\sqrt{2}$ is not a ratio is often attributed to Pythagoras himself, although it may have been found by one of his disciples.

IV. Zeno of Elea (c. 495–c. 430 B.C.) is best known for his paradoxes.

A. Zeno was interested in questions related to the *continuum*, that is, what it means to consider all possible distances and to assign a number to any possible distance. The idea that every single point between two distances corresponds to some number presents a paradox.

B. Consider an arrow in flight and its velocity (distance traveled divided by time). If every point between the beginning and the endpoint of the arrow's trajectory is a number, can we assign a velocity to the arrow when it reaches a specific point?
   1. Zeno believed that a velocity could not be assigned to the arrow because it doesn't travel any distance at any one instant in time.
   2. To get from the starting point to the endpoint, the arrow must pass over the continuum—but at each point on the continuum, the arrow is not moving.
   3. If at each point the arrow is not moving, how does it succeed in getting from where it is shot to where it lands?

V. Aristotle (384–322 B.C.) was one of the thinkers who attempted to circumvent Zeno's paradoxes by banishing them.

A. Aristotle also devised other separations that were significant for the development of mathematics. One of these is the difference between *arithmos* and *logos*.

B. *Arithmos* encompasses basic arithmetical calculations, such as those performed by people engaged in commerce or civil service.

C. *Logos*, which literally means "word," referred to closely reasoned, logical mathematical argument.

VI. Theaetetus of Athens (c. 417–369 B.C.) was associated with the Academy in Athens; his work is reflected in the writings of Euclid.

A. Theaetetus finished the description of the Platonic solids. These are solids whose faces are flat and are regular polygons. A cube is a simple example of a Platonic solid, each side of which is a square.

1. A tetrahedron is also a Platonic solid; it has 4 sides, each of which is an equilateral triangle. Other examples include the octahedron (8 triangles) and the dodecahedron (12 pentagons).
2. Theaetetus almost certainly discovered another Platonic solid, the icosahedron, which is made up of 20 equilateral triangles.

B. Theaetetus also worked on the recognition of rational numbers.

1. One of the problems encountered when working with rational numbers is that the same number can have very different appearances: $\frac{3}{4}=\frac{6}{8}=\frac{12}{16}$. We can see that these are all the same, but what if we had a more complicated fraction, such as $\frac{119}{91}$?
2. Theaetetus worked out a process known as the *Euclidean algorithm* for finding the greatest common divisor of two numbers. This process can be illustrated as follows:

- Start with $\frac{119}{91}$. Take out the largest integer, 1. The remainder is $\frac{28}{91}$.
- Take the reciprocal of that result, $\frac{91}{28}$.
- Take out the largest integer, 3. The remainder is $\frac{7}{28}$.
- Take the reciprocal of that result, $\frac{28}{7}$, which results in 4.

3. This process can also be reversed:

- Start with the last result, the integer 4. Take its reciprocal, $\frac{1}{4}$.
- Add the next integer, 3, to that result: $\frac{13}{4}$. Take its reciprocal, $\frac{4}{13}$.
- Add the next integer, 1, to that result: $\frac{17}{13}$, which is the original number, $\frac{119}{91}$, in reduced form.

4. Theaetetus realized that he could prove that $\sqrt{2}$ was not equal to a rational number by using the Euclidean algorithm combined with a geometric argument.

- Consider the length of a side of a square plus its diagonal: $1 + \sqrt{2}$.
- How many times does the side of the square go into that sum?
- That result is equal to 2 plus a number whose reciprocal is what we started with: $2+\cfrac{1}{2+\cfrac{1}{2+\cfrac{1}{2+\dots}}}$.
- This cannot be a rational number.

5. Working this process backward also gives an accurate method for approximating $\sqrt{2}$.

- Start with 2. Take the reciprocal and add 2: $2+\frac{1}{2}=\frac{5}{2}$.
- Take the reciprocal of that result and add 2: $2+\frac{2}{5}=\frac{12}{5}$.
- Continue this process to reach $\frac{169}{70}$. That result measures $1 + \sqrt{2}$ to an accuracy of 1 in 34,000.

VII. Eudoxus of Cnidus (b. c. 395–390 B.C.) was also a member of the Academy in Athens and interested in defining irrational numbers.

A. Theaetetus developed a procedure for determining that two rational numbers are equivalent to each other. Eudoxus worked out a procedure for determining that two irrational numbers are the same.

1. Eudoxus's procedure starts with two irrational numbers. If every rational number that is larger than one of these irrational numbers is also larger than the other one, and if every rational number that is smaller than one of these irrational numbers is also smaller than the other one, then the two irrational numbers must be the same.
2. Euclid picked up this idea and used it as a foundation for studying all numbers. He used the Euclidian algorithm for deciding whether two rational numbers are the same and Eudoxus's procedure for comparing the size of irrational numbers.

B. Eudoxus also calculated areas and volumes and is credited as the first mathematician to work out the *method of exhaustion*, which involves finding the area or volume of irregular regions or solids by dividing the region or solid into rectangular blocks.

VIII. Euclid of Alexandria, with whom we began this lecture, is a shadowy figure.

A. It is generally believed that Euclid lived roughly from 325 to 265 B.C., but in fact we do not have any definitive proof that he ever existed. No contemporary ever wrote about him, and the first biography was written 750 years after he lived.

B. However, the body of mathematical work that is attributed to Euclid, written around the year 300 B.C., was clearly written by one person; the writing is in the same voice with the same kinds of ideas and logical development.

C. Ptolemy I, the ruler of Egypt following the death of Alexander, built a center for scholars in Alexandria known as the Museion in the last years of the 4th century B.C. Euclid was almost certainly the leader of a group of mathematicians working at the Museion, tasked with collecting and explaining the mathematics known at the time.

1. We're not sure how much of the mathematics was original to Euclid, but his work is significant in the

logical foundation and explanations he set down for the discipline.

2. He made important choices about what terms needed to be defined and how they would be defined; what ideas needed to be taken as postulates, assumptions, or axioms; and how these would then be structured into proofs.
3. Euclid's *Elements* is still used as a textbook today. It deals with a number of important topics, including ratio and proportion, similarities in geometry, number theory, geometric progressions, prime numbers, irrational numbers, various kinds of magnitudes, areas and volumes, solid geometry, and the Platonic solids.
4. Euclid wrote other texts in addition to the *Elements*. We have some portions of two other books on geometry, *Data* and *On Divisions*, and his book on optics. He is said to have written other books that no longer exist, including one on surface loci (two-dimensional curves) and another with the intriguing title *Porisms*.

IX. In the next lecture, we will look at the next 700 years in Greek mathematics, from the time of Euclid until about A.D. 400.

Suggested Readings:

Euclid, *Elements*.

Heath, *A History of Greek Mathematics*, vol. 1.

Katz, *A History of Mathematics*, chap. 2.

Van der Waerden, *Science Awakening I*, chaps. 4–6.

Questions to Consider:

1. Why is it much easier to conceive of equality between two integers or two ratios of integers than between two irrational quantities? Why would this problem be so important to Greek mathematicians?
2. What was it about the Greek approach to mathematics that led them to establish a central role for proof?

Lecture Three—Transcript

Greek Mathematics—Thales to Euclid

In the last lecture, we looked at the early developments of mathematics in Mesopotamia and in Egypt. With this lecture, I am beginning a series of three lectures that look at Greek mathematics. In the Greek world, the Hellenic and Hellenistic worlds would dominate mathematical development for basically a thousand years—from about 600 B.C. until around A.D. 400.

I am going to focus in this lecture on the first 300 years—from about 600 B.C. right up until around 300 B.C. Something very important happens around the year 300 B.C., and that is when Euclid of Alexandria writes his great books. He is best known for the *Elements*, which is often considered to be a work of geometry. It actually has much more than geometry—it really covers much of the mathematics that was known at that time. Between the *Elements* and the other books that Euclid wrote, they really cover everything that was known in the Greek world—and for that reason, nobody bothered to keep any of the earlier Greek texts in mathematics. Euclid was so successful in summarizing what everybody else had done that there was no point in keeping what anybody else had done.

Incidentally, it is not clear how much of the mathematics and the writing of Euclid really was his own. There are some people who believe that he did not discover any mathematics by himself—that none of the theorems in Euclid's work really are original to him. There are other people who do argue that he must have been the first to come up with certain pieces of it. What is significant about Euclid's *Elements*, written around 300 B.C., is that they do summarize what was known before, and they set the foundation for everything that would come afterwards.

The earliest of the Greek mathematicians that we know something about is Thales of Miletus. He was born in the late 7th century and then would go into the 6th century. Thales was interested in a number of problems related to finding distances between objects. We know that Thales studied with the Egyptian mathematicians. He picked up much of their mathematics, and he did learn astronomy from them. It has been recorded that he predicted the eclipse that occurred in the year 585 B.C. It has also been recorded that he was able to calculate the height of the pyramids, and he would have used similar triangles in some way. It has also been recorded by Plato that Thales came up with a way of determining the distance that a ship is offshore. If you've got two different points on the shoreline, and you are able to observe the ship from both of these points, you can then do a geometric argument that enables you to determine the distance out to the ship. Plato also records one of the events that apparently happened to Thales, and it illustrates the concept of the absent-minded mathematician. Thales apparently one night was walking along studying the stars when he fell into a ditch, and a young servant girl who happened to be nearby pulled him out and asked him: How is it possible for you to understand what is going on up in the sky if you do not even see what is at your feet?

Thales of Miletus sets an important example for Greek mathematics because of his fascination with the idea of distance—calculating the heights of the pyramids, figuring distances out to a far ship. You are using geometric ideas, and certainly the other kinds of ideas of mathematics will appear within the Greek tradition. There is a lot that is built on problems of commerce, problems of taxation, as well as astronomy, but it is these geometric problems that really will come to dominate Greek mathematics. We are going to see that in a very large dependence on the idea of ratio and proportion—taking the ratio of two integers, one integer divided by another, and working with those ratios of numbers. Very often, instead of working with pure numbers, what the Greek mathematicians would do is to work with ratios.

We see this particularly strongly in the work of Pythagoras of Samos. Pythagoras was born sometime around 580 B.C. Samos is just a stone's throw from Miletus, and it is believed that at the very least, he probably met Thales. There is a tradition that he may actually have studied with Thales, but we do know that Pythagoras learned much of his mathematics by traveling to Egypt, and he actually was indoctrinated into many of the mystic practices of the Egyptian astrologer-priests. He learned a lot of the Egyptian astrology, or astronomy, or mathematics at this time. He happened to be in Egypt at the time that Cambyses of Persia conquered Egypt, and Pythagoras was taken back to Babylon by Cambyses, and he took advantage of this. There was still a very strong mathematical tradition in Babylon at this time—and according to the tradition,

Pythagoras learned a great deal of the Babylonian mathematics, the Babylonian astrology, or astronomy, at this time.

After leaving Mesopotamia and Babylon, Pythagoras went to Italy, and there he founded his own school—what became known as the Pythagorean School or sometimes the Pythagorean Cult, because he really adopted many of the cult practices, both of the Egyptian and also of the Babylonian or Mesopotamian mathematicians. He set up a group that was very mystical, very secretive, and we don't have many details of exactly what was going on in this group, but we do know certainly according to tradition that Pythagoras believed that mathematics lay at the core of all nature. The quote that Plato ascribes to him is: "At its deepest level, reality is mathematical in nature."

There certainly were cultish aspects of the Pythagoreans. They followed the Egyptian priests' practice of never eating beans and never wearing leather, but they also were clearly interested in developing the mathematics—and, in particular, developing mathematics built around ratio and proportion. And so music and the musical scale, which is very much built on ratios, is something that is generally attributed to the Pythagoreans. There are developments in architecture and the kind of proportions that you should have as you design buildings that are often attributed to the Pythagoreans.

Of course, the Pythagorean theorem is often attributed to them. As we know, the Pythagorean theorem is much older than the Pythagoreans—but, in fact, they appear to have been the first people to actually come up with the proof of the Pythagorean theorem. This was more than an observation—that the areas of those two squares on the legs, when you add those two areas, it is equal to the area of the square on the hypotenuse. According to tradition, it is the Pythagoreans who actually came up with the proof of this fact.

The Pythagoreans are also credited with coming up with the idea of an irrational number—that there are numbers that cannot be represented using fractions and cannot be represented using a ratio of two integers. The first example that we know of in this case is $\sqrt{2}$; $\sqrt{2}$ is about 7/5, but it is not exactly 7/5—and, in fact, you cannot find a fraction that is exactly equal to $\sqrt{2}$. We don't know when this was first proven. It is often attributed to Pythagoras himself, and one of the stories that is out there is Pythagoras discovered this fact, and there was a great feast among the Pythagoreans. Another one of the traditions is that it was one of Pythagoras's disciples who discovered this fact, and there was a great celebration that was experienced around this. The third tradition is that it was one of Pythagoras's disciples who discovered that $\sqrt{2}$ was not a ratio, that it could not be represented as a ratio of two integers, and Pythagoras was so disappointed and upset at this revelation that he banished this disciple from among the Pythagoreans.

The next mathematician I want to look at is Zeno of Elea. This is in what today is Italy. He is a 5th-century-B.C. mathematician, and he is best known for his paradoxes. He was interested in this whole question of what it means to consider all possible distances and assign a number to any possible distance. It is what today we refer to as the *continuum*. Does every single point between here and over there actually correspond to some number? If it does, then you start to get some paradoxes coming in. I want to focus on just one of them, because it is going to be important later on when we get to calculus. That is the problem if you consider an arrow in flight and you look at how fast it is traveling, its velocity, and you can determine the velocity by looking at how far it travels and how long it takes to get there: The distance that it travels divided by the time that it takes to get there is its velocity. If every single point between here and the endpoint of the arrow's trajectory is a number, we can think about what is happening to the arrow when it reaches a specific point. When it reaches a specific point in time or in distance, what is going on there—can we assign a velocity?

It seemed to Zeno that we cannot possibly assign a velocity, because if we are looking at the arrow at one instant, it doesn't travel any distance, and if it doesn't travel any distance then it can't possibly have a velocity. In any instant in time, the arrow is not moving, but to get from here to there we've got to pass over the continuum; we've got to pass over every single point—and at each point, the arrow is not moving. If at each point, the arrow is not moving, how does it succeed in getting from where I shoot it to where it lands? That is one of Zeno's paradoxes, and it would take a long time to really begin to grapple with this.

Aristotle, in the 4th century B.C., would be one of the people who attempted to work with Zeno's paradoxes by essentially banishing them. What Aristotle did was to separate mathematics into those numbers that are proportions—and deciding that we could work with them—and then the continuum that also contains these strange numbers like $\sqrt{2}$ that cannot be represented by a ratio, and simply put them into a whole other category.

There are other separations that Aristotle makes that turn out to be very significant for the development of mathematics, and one of those is the difference between *arithmos* and *logos*. *Arithmos* is the origin of our word "arithmetic," and so Aristotle is looking at basic arithmetical calculations—the kinds of calculations that would be used by people doing commerce or that would be used by the scribes and the civil servants—versus the kind of logical mathematical argument, the close reasoning that I talked about in the very first lecture and that he referred to as *logos*.

Logos is an interesting word. It literally means "word," and in the ancient Greek world, this word, *logos*, carried many, many different meanings. In fact, in the opening lines of St. John's Gospel, we have the word *logos* used: "*En arche en ho logos*" ("In the beginning was the Word"), but you could interpret that to refer to Aristotle's sense of *logos*: In the beginning was logical mathematical reasoning. I don't think that is quite what St. John had in mind when he wrote this, but it is interesting food for thought, and it certainly fits into much of the Greek tradition—and especially the tradition of the Pythagoreans.

The next mathematician that I want to look at is Theaetetus of Athens, born around 417 B.C. and died around 369 B.C. He was a friend of Plato. He was at the Academy in Athens. Theaetetus did a number of important things that would be reflected in the work of Euclid. In particular, Theaetetus is considered to have been the mathematician who finished the description of what today we call the *Platonic solids*. They are named for Plato, but they certainly do not come from Plato. These are solids that have faces that are flat, and the faces are all regular polygons. For example, a cube is a simple example of a Platonic solid, so each of the sides of a cube is a square, so a regular 4-sided polygon. There are other examples of Platonic solids. There is the tetrahedron, which has 4 sides; each of the sides of the tetrahedron is an equilateral triangle. Another example is the octahedron, which is made up of 8 triangles; and there is the dodecahedron, which is made up of 12 pentagons.

There is another Platonic solid. The tradition in Greek mathematics was that the Platonic solids came from the Pythagoreans, but almost certainly the last of these was not known to the Pythagoreans—almost certainly, it was originally discovered by Theaetetus—and that is the icosahedron, and this is a solid that is made up of 20 triangles, the faces of the solid are made up of equilateral triangles.

Something else that Theaetetus did was to work with the question: How do you recognize a rational number? One of the problems that we have when we are working with rational numbers is that you can have two very different appearances to the same rational number—thus, 3/4 is the same as 6/8, which is the same as 12/16. In those cases, it is easy to see that they are all the same fraction, but what if you've got a more complicated fraction, like 119/91? What is the simplest way of expressing that? It is not clear if this idea originally comes from Theaetetus, but he certainly exploited this, and this is the idea for thinking about the fraction 119/91: You first look at what is the largest integer in that, and the largest integer in that is 1. How much is left over? The remainder is 28/91. Once you take out the largest integer, whatever is left over is going to have to be a number between 0 and 1. Let's take its reciprocal. Instead of 28/91, we look at 91/28 and find the largest integer in that—28 goes into 91 three times, and what is left over is 7/28. Let's repeat that process and take the reciprocal of 7/28, and that is 28/7, and 7 goes exactly into 28 four times.

If you start out with a rational number, and you do this process—take out the largest integer, look at what is left, take its reciprocal, and take out the largest integer—that is going to stop after a finite number of times, and you can now redo this process. The integers that we got were 1, 3, and 4, and we can back up. Start with 4, and take its reciprocal and add 3 to that, and that gives us 13/4, and take its reciprocal, 4/13, and add 1 to that, and that gives us 17/13—and that gives us the original pair in their reduced form: 119/91 = 17/13. This is what would come to be known as the *Euclidian algorithm*.

What Theaetetus realized was that he could prove that $\sqrt{2}$ was not equal to a rational number by showing—and this is a geometric argument that unfortunately I don't have time to go into—but he used a geometric argument in which he showed that if you take the side of a square plus its diagonal and ask: How many times does the side of the square go into that? It goes in 2, with a little bit left over.

Geometrically, Theaetetus was able to show that if you take that little bit that is left over, and you look at its reciprocal, that is exactly the number of times that the side of a square goes into the side of the square plus its diagonal.

If we are trying to take the side of a square plus its diagonal, $1 + \sqrt{2}$, that is equal to 2 plus a number whose reciprocal is what you started with: $2+\cfrac{1}{2+\cfrac{1}{2+\cfrac{1}{2+\ldots}}}$. This continues forever. It never terminates.

This cannot be a rational number. Incidentally, this gives you a very accurate way of approximating $\sqrt{2}$, and that is to work this process backwards. Start with 2. Take its reciprocal and add 2: $2 + 1/2 = 5/2$. Take that reciprocal and add 2 to it—that gives you 12/5. Take its reciprocal, 5/12, and add 2 to that—that gives you 29/12. Take its reciprocal and add 2 to that, and that gives you 70/29. Take its reciprocal and add 2 to that, and you get 169/70, and I hope you see how you can continue this as long as you want and get as accurate as you want. But if I take 169/70, that accurately measures $1 + \sqrt{2}$ to an accuracy within 1 in 34,000, and you can quickly get whatever accuracy actually is needed.

The next Greek mathematician I want to look at is Eudoxus of Cnidus—also someone who was at the Academy in Athens. He was born sometime between 395 and 390 B.C. Eudoxus of Cnidus was interested in this question: How do you define an irrational number? If you've got a number like $\sqrt{2}$, we have seen how Theaetetus was able to take the problem of when are two rational numbers equivalent to each other and work out a procedure for deciding that. What we are looking at here with Eudoxus of Cnidus is how do you decide if two irrational numbers are the same, and he came up with a brilliant idea that actually would be adopted in the 19th century of the modern era by Richard Dedekind, who we will be talking about much later.

The way to compare two irrational numbers and decide if they really are the same number is to look at the rational numbers that are larger than each of them, and if every rational number that is larger than this irrational is also larger than the other one, and if every irrational number that is smaller than this is also smaller than the other one, then the two irrational numbers have to be the same. This is an idea that Euclid would pick up and he would use as an important foundational aspect for studying all numbers—whether they are rational numbers, in which case he would explain what we call the Euclidian algorithm for deciding when two rational numbers are the same—or for comparing the size of irrational numbers.

Eudoxus of Cnidus is also important for his work on areas and volumes, and he is credited with being the first mathematician to really work out how to take this problem of finding the area of an irregular region or taking the volume of something that is not just a rectangular solid and approximating the area by chopping it up into little rectangular blocks. This is a method that would come to be known as the *method of exhaustion.* Euclid describes it in his writings, and in the next lecture I am going to talk about how Archimedes of Syracuse really refined this idea and turned it into a very powerful way of working with the problem of finding areas and volumes.

Euclid of Alexandria is a shadowy figure. It is generally believed that he lived roughly from 325 to 265 B.C., but the fact is not only do we not know when he was born or when he died, we don't even have any definitive proof that he ever existed. There is no contemporary who wrote about him. There is a reference in one of the works of Archimedes, but a lot of the scholars believe that that was a later edition. We don't get the first biography of Euclid until 750 years after he lived. However, there is a body of mathematical work that was written around the year 300 B.C. that clearly was all written by the same person; it is in the same voice, with the same kinds of ideas and the same logical development. Somebody must have done this writing, and we might as well believe that it is the person to whom it is attributed, this Euclid of Alexandria.

Alexandria is a city that had been founded after the death of Alexander, the first ruler of Egypt. Following the death of Alexander was Ptolemy I, and he built a center for scholars—what was known as the *Museion*. This was a temple to the muses. It is the origin of our word today "museum." The Museion was set up by Ptolemy I, in the last years of the 4th century B.C., to be a center for scientific work and for scholarly work. There was a group of mathematicians who were pulled together at the Museion, and Euclid almost certainly was the person who was leading this group, and his job was to take the mathematics that was around him and condense it and explain it.

While we are not sure how much of this mathematics was original to Euclid, what was very original to Euclid was the way in which he was able to explain the mathematics and put it on a very logical foundation. Euclid clearly made important choices about what needed to be defined and how it would be defined; what needed to be taken as a postulate, or assumption, or axiom; and how these then would be structured into proofs. Euclid was one of the greatest teachers to have ever existed.

Euclid should be thought of as a great textbook writer, and that is really what is going on here. He is taking ideas, most of which existed in Greek mathematics before his time, and he is organizing them in an incredibly effective way. Euclid's *Elements* is still a textbook that is used today. It has gone through well over a thousand editions in many, many translations, and it is made up of many books. Usually when you think of Euclid's *Elements*, you think of geometry, and the first four books of Euclid's *Elements* really are devoted to geometry, but there is much more mathematics going on here in addition to that. Books 5 and 6 deal with ratio and proportion. Those books are also looking at similarities in geometry. Books 7 through 9 look at number theory, and here is where Euclid explains this Euclidian algorithm that really comes to us from Theaetetus or possibly even earlier than that.

Euclid looks at sums of geometric progressions. This is where Euclid proves that there are infinitely many prime numbers. A prime number is one that is only divisible by 1 and itself among the positive integers. Euclid proves that, in fact, there are infinitely many such primes: 2, 3, 5, 7, 11, 17, and so on. Book 10 of the *Elements* looks at the meaning of irrational numbers, and here is where he picks up this idea of Eudoxus of Cnidus, on how to define an irrational number, and Book 10 talks about how to compare various kinds of magnitudes.

Chapters 11 through 13 look at solid geometry. He explains Eudoxus's method of exhaustion and how to find areas and volumes of irregular objects. He explains how to construct the Platonic solids, and he goes well beyond that. He doesn't just explain what the Platonic solids are, he actually proves that these are the only ways that you can construct a solid so that all of the faces look exactly the same and all of the faces are, in fact, regular polygons—polygons in which all of the sides have exactly the same length.

There is much else that Euclid wrote. He also wrote two other books on geometry. We still have some pieces of those—one called *Data* and one called *On Divisions*. He wrote one of the first books on optics, and in that book we find the very first recorded treatment of the question of perspective. Another one of Euclid's books is a book on astronomy called *Phenomena*, and there are several other books that people recorded that Euclid had written but unfortunately that no longer exist. There is one book on surface loci (so this would be two-dimensional curves). There is another book with an intriguing title—*Porisms*—and people are not quite sure what is meant by a *porism*, but from some of the examples that other people make of the kinds of results that Euclid stated in this book, these appear to be results that can easily be deduced from other theorems that he proved. Most of the porisms are geometric results that follow from the main theorems that Euclid proved in the *Elements*. There was also a book on conics, and there is also a book called *The Book of Fallacies*, and he had at least one book on music.

For the next time, we are going to look at the following 700 years of Greek mathematics—from the time of Euclid right up until about A.D. 400. We are going to focus in particular on Archimedes, and we will also be looking at Apollonius, Diophantus, and the first recorded woman in mathematics: Hypatia of Alexandria.

Lecture Four

Greek Mathematics—Archimedes to Hypatia

Scope:

This lecture begins with Eratosthenes, who worked in the Museion established by Ptolemy I. Among his many accomplishments was the determination of a method for calculating the circumference of the Earth. From Eratosthenes, we move on to the greatest of all Hellenistic mathematicians, Archimedes of Syracuse. We'll look at two examples of his mathematics—a method for calculating the volume of a sphere and the computation of π to arbitrary precision. Another Hellenistic mathematician, Apollonius of Perga, wrote a complex and sophisticated text on conic sections that would prove to be indispensable to Isaac Newton almost 2000 years later. The last of the great Hellenistic mathematicians was Diophantus of Alexandria, who was the first person known to use a single letter to represent an unknown quantity, as we do in algebra today. We close the lecture with Hypatia of Alexandria, the first woman recorded to have made important contributions to mathematics.

Outline

I. This lecture picks up after Euclid and continues with Hellenistic mathematics to about A.D. 400. In the next lecture, we'll look at the astronomical work being done in the same period.

II. As mentioned in the last lecture, Ptolemy I established the Museion in Alexandria, which would become a center for mathematical work for the next 700 years. His successors, Ptolemy II and Ptolemy III, founded and stocked the Library of Alexandria, the greatest repository of books and scrolls in the ancient world.

- **A.** One of the first mathematicians to follow Euclid in the Museion was Eratosthenes (276–194 B.C.), originally from Cyrene. Among his accomplishments was the discovery of a general method for finding all the primes below a given number. His method is known as the *sieve of Eratosthenes*, and computers searching for primes today rely on the idea behind his method.

B. Eratosthenes also measured the circumference of the Earth, which was known to be round. Observations of ships disappearing over the curve of the horizon and the shadow of the Earth projected onto the Moon during a lunar eclipse would have shown the shape of the planet. Eratosthenes devised a method for determining its circumference.

- Find a place where the Sun is directly overhead at a certain time of year.
- Then choose another place a certain distance north of the first location.
- Determine the angle made by the Sun's rays at the second point, and use this angle to determine the total circumference.

1. The Sun is so far away from the Earth that its rays hit the Earth essentially as parallel lines. The total circumference of the Earth can be determined using this knowledge.
2. Eratosthenes performed this calculation by digging two wells, one in southern Egypt and one near Alexandria. When the Sun was directly above the well in southern Egypt, he looked at the angle made by the rays at the well near Alexandria. He then accurately measured the distance between the two wells and determined the circumference of the Earth to be between 24,000 and 25,000 miles.

III. Archimedes of Syracuse (287–212 B.C.) was a colleague of Eratosthenes and the greatest of all the mathematicians in the Greek world.

A. Archimedes was known as an engineer and is credited with inventing the hydraulic screw. He was also instrumental in developing the idea of the block and tackle and discovered the power of levers. He famously said, "Give me a place to stand and I will move the Earth."

B. Archimedes turned his attention to machines of war, particularly catapults, as well as optics and the properties of mirrors. According to tradition, in 214 B.C., when the Roman General Marcellus was attacking Syracuse,

Archimedes instructed the soldiers defending the city to line up with polished mirrors, focus them on Marcellus's ships, and set the ships ablaze.

C. The inventiveness of Archimedes was well-known to the Romans, and it was said that the Roman soldiers were terrified of his engines of war.

1. Marcellus eventually conquered Syracuse in 212 B.C. and, according to legend, instructed one of his soldiers to find Archimedes and bring him safely back to the Roman camp.
2. When the soldier found the great mathematician amid the chaos of the conquered city, Archimedes was deep in performing mathematical calculations on a sandboard.
3. The soldier ordered Archimedes to return with him to the camp, but when Archimedes said that he had to finish his calculations first, the soldier killed him.

D. Archimedes had asked that his tomb feature a sphere contained within a cylinder to commemorate his calculation of the volume of a sphere.

1. Archimedes's formula is that the volume of a sphere is $\frac{2}{3}$ the volume of a cylinder. This can be connected to the usual formula, $\frac{4}{3}\pi r^3$, as follows:

- Place a sphere inside a cylinder so that the circumference of the sphere exactly matches up with the circumference of the cylinder and the height of the sphere is equal to the height of the cylinder.
- The volume of the cylinder is the area of the base (πr^2) times the height of the cylinder, which is the diameter of the sphere $(2r)$. This produces the formula $2\pi r^3$.
- If the sphere is $\frac{2}{3}$ the volume of the cylinder, then the sphere is $\frac{2}{3} \times 2\pi r^3$, which is $\frac{4}{3}\pi r^3$.

2. Archimedes worked this formula out using the method of exhaustion developed by Eudoxus of Cnidus. In other words, he broke the volume of the sphere into thin slices

and then added the volumes of all the slices to approximate the volume of the sphere.

3. This same idea is behind integral calculus, which at its heart deals with the problem of finding areas and volumes, and which works by this slicing mechanism.

E. Archimedes also found a way of approximating π, essentially to arbitrarily high accuracy.

1. The number π is the number of times that the diameter of a circle is used to measure off the circumference of the circle. The diameter goes around the circumference a little bit more than 3 times.
2. Archimedes began by considering a hexagon, a 6-sided figure made up of equilateral triangles that fits precisely into a circle. The circumference of the hexagon is 6 times the radius of the circle, or 3 times the diameter.
3. Archimedes reasoned that with a regular polygon inside of a circle, he could double the number of sides of the polygon and find the relationship between the new circumference and the old circumference.
4. Once he knew that a hexagon inside a circle has a circumference of 3 diameters, he was able to work out the circumference of a regular polygon with 12 sides, 24 sides, 48 sides, and 96 sides. Thus, he was able to measure π accurately to about three digits, or about 3.14.

IV. All of our evidence about work with conic sections in Hellenistic times comes from Apollonius of Perga (b. c. 260 B.C.).

A. Conic sections are curves obtained by slicing a pair of cones that are matched point to point.

1. If we make a horizontal slice through one of the cones, we get a circle. If we tip the plane that slices through the cone slightly, we get an ellipse. If we tilt the plane until it is actually parallel to one of the sides of the cone, the ellipse becomes a parabola.
2. If we continue to tilt the plane, it intersects the other cone, and we get one piece from the bottom cone and one piece from the upper cone. Those two pieces together give us a curve called a *hyperbola*.
3. All these basic curves, the circle, ellipse, parabola, and hyperbola, are united in this idea of slicing cones, and

Apollonius explained the properties of these conic sections.

B. Apollonius's work on conic sections would eventually become an important part of analytic geometry and was used by Isaac Newton to explain the motion of the planets around the Sun. Today, we find conic sections easier to understand algebraically.

V. A number of important innovations come to us from Diophantus of Alexandria (c. A.D. 200–284).

A. Diophantus was the first person known to use a single letter to represent an unknown quantity, such as the variable x we use in algebra. He was known to Islamic mathematicians, the inventors of algebra, and he was responsible for a compact notation to represent quadratic polynomials—that is, polynomials that involve the square of an unknown plus a linear term plus a constant $(x^2 + bx + c)$—and cubic polynomials.

B. Diophantus is best known for his book *Arithmetica*, in which he studies problems with solutions that are only integers. Today, we call these *Diophantine equations*; classic examples of such equations are the Pythagorean triples.

1. If we have the equation $x^2 + y^2 = z^2$, and we let x and y stand for any positive numbers, we can always find a positive value for z.
2. The more interesting question is: Can we find integers for x and y that lead to an integer value for z? Solutions include those we saw earlier: (3, 4, 5), (5, 12, 13), and so on.
3. Diophantus was the first person to record a method for generating all the possible Pythagorean triples, although the Babylonians probably knew this method.

VI. The first woman to appear in the history of mathematics is Hypatia of Alexandria (c. A.D. 370–415).

A. Hypatia was the daughter of one of the scholars associated with the Museion, Theon of Alexandria. She was an accomplished mathematician and tutor. Although we have no original mathematics directly attributable to Hypatia, we know that she wrote commentaries on the work of

Archimedes, Ptolemy, and Diophantus. In 1968, an Arabic translation of commentaries on Diophantus's *Arithmetica* was discovered, which is believed to have been written by Hypatia.

B. The Museion came to an end during the life of Hypatia. The last references to this center for scholarship occur in the late 4th century and probably coincide with the banning of all pagan temples by the Christian Emperor Theodosius I in A.D. 391.

C. Hypatia continued to live and work in Alexandria, but the arrival of the Christians ended the great period of mathematical achievement there.

1. Alexandria became embroiled in a power struggle for control of the city between Orestes, the prefect of Alexandria and the emperor's representative, and Cyril, the Christian bishop of the city.
2. Hypatia, a strong supporter of Orestes, was attacked and torn to pieces by a Christian mob on the streets of Alexandria in A.D. 415.

VII. In the next lecture, we'll turn to the development of astronomy in the Hellenistic world.

Suggested Readings:

Dijksterhuis, *Archimedes*.

Heath, *A History of Greek Mathematics*, vol. 2.

Katz, *A History of Mathematics*, chaps. 3, 5.

Stein, *Archimedes: What Did He Do Besides Cry Eureka?*

Van der Waerden, *Science Awakening I*, chaps. 7, 8.

Questions to Consider:

1. For Archimedes, would there have been practical uses for a highly accurate approximation to π? Why do you think he explored π to such a high degree of accuracy?
2. Diophantus was the first to use a single letter to represent an unknown quantity. Does this constitute the invention of algebra? If not, what more would be needed?

Lecture Four—Transcript
Greek Mathematics—Archimedes to Hypatia

In this lecture we are going to be looking at Hellenistic mathematics. The last lecture ended with Euclid of Alexandria, sometime early in the 3rd century B.C. We are going to pick up there and continue right through until A.D. 400. This lecture and the next lecture are going to be separated. In the next lecture I want to put all of the work on astronomy, and so this lecture is going to focus on 700 years of Hellenistic mathematics, but without any mention of the astronomical work that would be done.

As I mentioned in the last lecture, it was Ptolemy I who established the Museion in Alexandria, in what today is Egypt, and this would become a center for mathematical work actually for the next 700 years. Aside from Ujjain in India, there probably is no other place on Earth that has been a center of important mathematical activity for such a long stretch of time.

Ptolemy's successor, Ptolemy II, would go on to build a great library that would be there to support the Museion—and his successor, Ptolemy III, then would make a point of really building that library up into the greatest repository of books anywhere in the world. In fact, there are many stories of people being pressed into loaning copies of their manuscripts so that they could be recopied. Anyone who traveled to Alexandria and was taking a book or scroll with them was required to loan this scroll to the library, and sometimes if it was a very valuable scroll, the scroll would not be returned to the person that it was taken from. Rather, they would have to leave with a copy of the book that they arrived with. We also know that Ptolemy III did get in contact with many other rulers of the eastern Mediterranean and aggressively went after books from other sources.

One of the first mathematicians to follow Euclid in the Museion was Eratosthenes—originally from Cyrene, born around 276 B.C.; he would die around 194 B.C. In addition to being a mathematician, he was the third librarian of the great library in Alexandria, and he did a number of very interesting mathematical accomplishments—one of which is to find a general method for finding all of the primes below a given number. For example, if you want a list of all of the primes less than a million, there are about 78,000 of them. There is an algorithm, or procedure, that Eratosthenes came up with called the *sieve of Eratosthenes*, which is a method of generating all of the primes—and even today, with our powerful computers, the kind of procedure that these computers use in order to find all of the primes below a given number, such as the primes less than a million, rely on the idea behind Eratosthenes's sieve.

Something else that Eratosthenes accomplished was to measure the circumference of the Earth. It was known in ancient times that, in fact, the Earth is round. It is a common myth that people right up into the Middle Ages believed that the Earth was flat—in fact, they were quite aware of the fact that it was round. One of the arguments for knowing that it is round is the fact that as a ship sails out to sea, you see the bottom of the ship disappear first, and the mast is the last part of the ship to disappear because it is disappearing over the curve of the horizon—but there is an even more immediate way of seeing that the Earth must be round, and that is by looking at a lunar eclipse. A lunar eclipse happens when the Earth comes between the Sun and the Moon, and so the shadow of the Earth is projected onto the Moon, and you only have to look at that shadow to see that it is round. The Earth has to be round. If it is round, then the question comes up: How large is the Earth? This is a question that Eratosthenes wrestled with, and he came up with a very ingenious way of figuring it out.

It is known that the Sun is so far away from the Earth that the rays of the Sun hit the Earth essentially as parallel lines, and so what you can do is find a place where the Sun is directly overhead at a certain time of year—and then find another place, a certain distance north of that, and figure out the angle that the Sun's ray makes at that northernmost point, and compare that angle to figure out that distance, what angle of the total circumference of the Earth is being measured here, and use that to extrapolate the circumference of the Earth.

Eratosthenes actually did this. What he did was to dig wells—one in southern Egypt and another one near Alexandria. He chose a time when the Sun was directly above the well in southern Egypt, and he looked at the shadow, the angle that the Sun made with the well near Alexandria. He was very lucky in being able to accurately measure the distance between the two wells, and he came up with a measurement of the circumference of the Earth that essentially is right on. He came up with something between 24,000 and 25,000 miles, or the equivalent of the units of that time, of something between 24,000 and 25,000 miles, which is what the circumference of the Earth actually is.

One of the problems, of course, is that there were lots of people who were estimating the circumference of the Earth at this time. Eratosthenes's estimate was one of many estimates that were out there, and an independent observer didn't have any good way of knowing which one was correct. It wouldn't be known until much later that, in fact, Eratosthenes had been correct.

I want to spend most of my time in this lecture looking at one of the friends, and colleagues, and contemporaries of Eratosthenes—this is Archimedes of Syracuse, certainly the greatest of all of the mathematicians of the Greek world. He was born around 287 B.C. and died in 212 B.C. He was noted as an engineer. He is credited with inventing the hydraulic screw. We know that he went to Egypt, that he did spend some time in Alexandria, and the tradition there is that he invented the hydraulic screw—this is a screw mechanism that can be used to lift water—and he invented it for use in Egypt. He probably did not invent the block and tackle, but he certainly understood the principles behind the block and tackle, and he was very instrumental in developing this idea. He also discovered the power of levers, and he is famous for his quote: "Give me a place to stand and I will move the Earth." If you've got a fulcrum point, and you've got a long-enough lever, somebody who is as puny as a mere human being is actually capable of moving an object as large as the Earth through the mechanical advantage of using a lever.

Archimedes also turned his attention to vehicles and machines of war, catapults in particular. Burning mirrors are something that he is credited with inventing. Certainly Archimedes was very much involved with optics, with the properties of mirrors. The tradition is that in 214 B.C., when Marcellus, the Roman general, was attacking Syracuse, one of the defenses of the city that Archimedes came up with was to line up soldiers with polished mirrors that he would focus on the ships and set them ablaze. Almost certainly this didn't happen, and an interesting item of debate today is whether this is even possible. Some people argue that it is not possible; others argue that if he got things just right, there are ways in which he might have been able to succeed.

One of the important points is that Archimedes was well-known to the Romans as somebody who was extremely inventive. It was said that the Roman soldiers were so terrified of these engines of war that Archimedes was inventing, that all the defenders of Syracuse had to do was to drape a little piece of rope over the parapet, and the Roman soldiers would assume that Archimedes was about to unleash some terrible new weapon, and they would flee in terror.

The Romans would eventually conquer Syracuse in 212 B.C.—and it was again the Roman general, Marcellus, who was leading the Romans when they did conquer Syracuse. Marcellus was well aware of Archimedes, and he greatly valued the contributions that Archimedes was making, even if he was working for the people in Syracuse who were his enemy.

The story is that Marcellus sent one of the soldiers to go and find Archimedes and bring him safely back to the Roman camp, and that the Roman soldier misunderstood the order, simply that he was supposed to bring Archimedes back. The story is told that the soldier came to the house of Archimedes, general chaos going on in the streets around him as the soldiers were entering the city and pillaging and burning the city. The soldier came up, and he discovered Archimedes deep in a mathematical piece of work. There is a mosaic that shows this. Archimedes was doing his calculations on a sandboard, which is probably the way he really was doing calculations at that time. You wouldn't want to waste valuable papyrus or other writing surface in order to put something indelible if you were just doing mathematical calculations, and so you would write them with your finger on the board that was covered with fine sand; once you found your answer, then you could wipe it out and reuse the sand board again.

The soldier finds Archimedes deep in his mathematical calculations, and he orders him to follow him, and Archimedes says: No, let me finish these calculations first. And the soldier—who believes that he must get Archimedes back without delay—kills him on the spot, rather unfortunately. And so that is the death of Archimedes.

I have described Archimedes as a great engineer, and certainly that is where he made much of his reputation at the time, but he also left a very important mathematical legacy. In fact, we've got an indication that it was the mathematics that was most important to Archimedes, because Archimedes had asked that on his tomb there be erected a sphere contained within a cylinder, and this refers to Archimedes's calculation of the volume of a sphere, which he considered to be his greatest work. In fact, we do have independent confirmation Cicero visited Syracuse in the year 75 B.C., and he found the cemetery where Archimedes was supposed to be buried—and in fact, he did find this tomb of a sphere inside a cylinder. The result that Archimedes found is not the usual way we have of stating the formula for the volume of a sphere. You remember that formula if you have been taught it in grade school, as $(4/3)\pi r^3$—so, 4/3 times π times the cube of the radius.

The way Archimedes stated this was that the volume of the sphere is 2/3 the volume of the cylinder, so if we put a sphere inside a cylinder so that the circumference of the sphere exactly matches up with the circumference of the cylinder, and the height of the sphere is equal to the height of the cylinder, it is easy to work out the volume of the cylinder. The volume of the cylinder is the area of the base, which is πr^2, times the height of the cylinder, which is the diameter of the sphere, which is 2 times r ($2r$). You take the base, πr^2, times $2r$, and you get $2\pi r^3$ for the volume of the cylinder. If the sphere is 2/3 the volume of the cylinder, then the sphere is $(2/3)2\pi r^3$, which is $(4/3)\pi r^3$. The way the ancients thought about the volume of the sphere is that it was 2/3 the volume of the cylinder.

Archimedes worked this out. He worked it out using the method of Eudoxus of Cnidus, this method of exhaustion that Euclid had talked about: Taking this volume and breaking it down into thin slices, you can find the volume of each of the thin slices and then approximate the volume of the entire sphere by adding up the volumes of the thin slices and seeing what that number approaches as you take progressively thinner slices of the sphere.

This is the idea behind integral calculus. Integral calculus, really at its heart, has the problem of finding areas and volumes, and it works by precisely this kind of slicing mechanism that was first introduced by Eudoxus, and later developed by Euclid, and really perfected by Archimedes. It is not too much of a stretch to say that Archimedes understood integral calculus. When we get to the 17th century, we will see that all of calculus involves much more than being able to find areas and volumes, but this basic idea is something that Archimedes really understood.

Something else that he was able to do is to find a way of approximating π, essentially to arbitrarily high accuracy. Remember that π is the number of times that the diameter of a circle is used to measure off the circumference of the circle. The diameter goes around the circumference 3 and a little bit, and the number of times the diameter goes around the circumference is this number we refer to as π. Incidentally, the use of that particular letter, the Greek letter π, for *perimetros*, is something that would be introduced in the 18th century of the modern era. Before then people did not use this Greek letter π to represent the ratio of the circumference to the diameter of a circle.

Nevertheless, people were interested in this ratio, and Archimedes began by considering a hexagon, a 6-sided figure, that fits precisely into a circle. If I've got a hexagon, a 6-sided figure, it is made up of 6 equilateral triangles, and that means that the circumference of the hexagon is 6 times the radius of the circle—or in other words, it is 3 times the diameter. If I start out with a 6-sided figure (a regular hexagon) in my circle, its circumference is going to be exactly 3.

What Archimedes succeeded in doing is figuring out if I've got a polygon that sits inside a circle, a regular polygon, all of the sides have the same length, and he figured out how to take that polygon and double the number of sides and find the relationship between the new circumference and the old circumference. Once he knew that a hexagon inside this circle has a circumference of exactly 3, he was able to work out the circumference of a 12-sided regular polygon. From there, to 24 sides, and from there to 48 sides, and from there to 96 sides—and he was able to measure π accurately to about three digits, about 3.14, which comes out of recognizing that π can be approximated with this 96-sided regular polygon. Archimedes could have gone further with that, but he chose to stop at that point.

The next mathematician I want to talk about is Apollonius of Perga, born sometime around 260 B.C., and the big contribution that I am going to talk about today from Apollonius is his work on conic sections. As I mentioned in the last lecture, we know that Euclid wrote a book on conic sections, but we don't know what he said about them, and all of our evidence of what was developed in Hellenistic times about conic sections comes from Apollonius.

Incidentally, we are going to encounter Apollonius again in the next lecture because he also has important contributions that he made to astronomy, but what he was looking at was conic sections. These are curves that can be obtained by slicing a pair of cones. I take a pair of cones that meet point to point, and I consider what happens as I slice through these cones. If I make a horizontal slice through one of the cones, I am going to get a circle. If I tip that plane that slices through the cone slightly, it is going to elongate my circle, and I am going to get an ellipse—so a circle that is simply elongated along the axis, in the direction in which I am slicing by this plane.

As I move the plane up to an ever-steeper angle, my ellipse gets longer and longer and longer, until finally I take a plane that is actually parallel to one of the sides of the cone—so that the plane comes in and hits the cone and then never leaves, and what happens is that my ellipse stretches out so far, and then suddenly it opens up at one end, and I get a parabola.

If I now continue tipping that slicing plane, if I tip it just a little bit more, it is now going to intersect the other cone, and as it intersects the other cone, I am going to get two pieces. I am going to get one piece on the bottom cone and another piece on the upper cone, and those two pieces together give us the curve that is called a *hyperbola.* We've got the circle, the ellipse, the parabola, the hyperbola—and all of these basic curves are united in this idea of slicing cones. What Apollonius of Perga is able to do is to explain properties of these conic sections, these various curves that are all related in this way, and find the things that are common among them.

Apollonius's work on conic sections would eventually become an important part of analytic geometry, when geometry and algebra get merged in the 17th century. Today when we study these conic sections, we usually think about them algebraically, and it makes it much, much easier to understand conic sections if you think about them algebraically. Apollonius did not have this kind of algebraic representation at hand. He had to think of them purely geometrically, and it is incredible how much he was able to do.

We know that Isaac Newton, as he was writing the *Mathematical Principles of Natural Philosophy*, explaining the motion of the planets around the Sun, he needed to use conic sections, because as was discovered by Kepler, the planets move in conic sections—they move in ellipses, or parabolas, or hyperbolas. We know that Newton directly drew on this work of Apollonius of Perga in order to do his work in the *Mathematical Principles of Natural Philosophy* in the late 17th century of the modern era.

Now jumping ahead several hundred years, because I am going to jump over some important astronomers who will figure in the next lecture, the next mathematician I want to look at is Diophantus of Alexandria, born sometime around A.D. 200 and died about A.D. 284—certainly a mathematician in Alexandria of the 3rd century A.D.

A number of important innovations that come to us from Diophantus—he is the first person known to use a single letter to stand in for an unknown quantity. Today in algebra we use x all the time, or t, or some other letter to represent the unknown, and it was Diophantus of Alexandria who—so far as we know—was the first person to come up with this idea. That does not mean that he invented algebra—as we will see especially when we get to the lecture on Islamic mathematics, there is much more to algebra than just having a single letter that represents an unknown, or even working with things like quadratic polynomials, which we saw that the Babylonians were able to do.

But Diophantus did help set the stage for the later development of algebra. He was very well-known to the Islamic mathematicians. He is responsible for a very compact notation to represent quadratic polynomials—polynomials that involve the unknown squared, plus a linear term, plus a constant. Diophantus also studied cubic polynomials—these are polynomials that go up to a cubed term, something raised to the third power. Diophantus of Alexandria is best known for his book *Arithmetica*, in which he studies problems where he is only looking for solutions that only involve integers—what today we call *Diophantine equations*, in honor in Diophantus.

A classic example of this is the Pythagorean triples that I talked about in connection with Babylonian mathematics. So you've got an equation like $x^2 + y^2 = z^2$, and you can let x and y be any positive numbers, and there will be a positive value for z—but the more interesting question is: Can you find integers for x and y that lead to an integer value for z? Examples like (3, 4, 5), (5, 12, 13), and so on. It was Diophantus who studied this question, and Diophantus is the first person who actually recorded a method for arbitrarily generating many of these—and, in fact, all of the possible Pythagorean triples.

Going back to the Babylonian work from almost 2000 B.C., we have seen that they were able to generate literally dozens of these Pythagorean triples. They probably knew the method that Diophantus explained in the 3rd century A.D., but nobody actually wrote it down until we got to Diophantus.

The last mathematician that I want to talk about is the first woman to appear in the history of mathematics, and that is Hypatia of Alexandria. She was born around A.D. 370. We know quite precisely when she died. She died in A.D. 415. She was not alone in being a woman who was doing mathematics.

We do have accounts of the mathematical community surrounding the Museion at this time in Alexandria, toward the end of the 4th century A.D. We know that the Museion itself was made up of a group of scholars. Many people would come from around the Hellenistic world in order to study mathematics, study mechanics, study optics, study the other sciences there at the Museion with these great scholars. The great scholars incidentally—so far as we know—were not paid, but they were fed, and that was good enough to get them all to come together. We also know that the Museion had a number of lecture halls where the members of the Museion could lecture about their topic. People would come, and they would listen to these lectures, and they probably would pay the scholars independently for further lessons. There was a group of people who gathered around the Museion and who would also offer private tutoring or other lessons in mathematics, and Hypatia was one of them. She was actually the daughter of a member of the Museion, Theon of Alexandria.

She was quite accomplished. We know that while there is no original mathematics that is directly attributable to Hypatia, we know that she wrote commentaries on the work of Archimedes and the work of Ptolemy, the astronomer who I will be talking about in the next lecture. She wrote commentary on the work of Diophantus. In fact, just recently we discovered a work that is believed to have come from Hypatia. In 1968—there was an Arabic translation of commentaries written on Diophantus's *Arithmetica*, and it is generally believed from looking at the contextual evidence that these commentaries that were rediscovered in 1968 probably were written by Hypatia.

If there was no original mathematics—and sometimes that is hard to tell because the commentaries would often elaborate on the mathematical things and develop them further—certainly she was someone who understood the classic texts so well that she could write commentaries that would be used for hundreds and, in fact, well over a thousand years later.

It is under Hypatia that the Museion comes to an end, and we do not know the exact date or the exact circumstances, but the last references to the Museion occur in the late 4th century, and they probably coincide with an event that happened in A.D. 391. That is the year in which the Roman Emperor, Theodosius I, who was a Christian emperor, banned all of the pagan temples. He ordered all pagan temples to be closed throughout the Roman Empire. The Museion was a center for scholarship, but remember that it was founded as a temple to the muses, and as such it fell under this ban of Theodosius I, and so almost certainly at this point the Museion is suppressed, and the scholars who would have gathered in this temple to the muses were disbanded.

Hypatia continued to live in Alexandria and to work in Alexandria, but the coming of the Christians to Alexandria would really mark an end to this great period of mathematical achievement. In A.D. 415, there is a great political fight between two leaders. On the one hand we have Orestes—who was the prefect of Alexandria, the emperor's right hand man—running Alexandria, and on the other side we have Cyril, who was bishop of Alexandria, the leader of the Christians in Alexandria. And since the emperor himself was a Christian, to be the bishop of Alexandria was to have a great deal of power. Cyril and Orestes were engaged in a battle to see who could really control the city of Alexandria—and with control of the city of Alexandria would really come control of Egypt. We know that Hypatia was a strong supporter of Orestes—against the Christian bishop, Cyril. In the year A.D. 415, she was walking through the city of Alexandria when a Christian mob spotted her and attacked her, and she was literally torn to pieces. And that marks the last of the great Hellenistic mathematicians and really the end of the Hellenistic period in mathematics.

We are going to be looking at the development of astronomy in the Hellenistic world, and then in the following lecture we are going to move to India, South Asia, and we are going to see how they were able to pick up the Hellenistic tradition and further develop it.

Lecture Five

Astronomy and the Origins of Trigonometry

Scope:

Trigonometry was not applied to problems of surveying until the very end of the 16th century A.D. Its origins lie in problems of astronomy—in particular, the question of why the seasons are not the same length. Hipparchus of Rhodes, considered the father of trigonometry, answered this question by discovering a method for determining the length of the chord that connects the endpoints of a given arc of a circle. His methods were greatly refined by Ptolemy of Alexandria, for whom trigonometric relations and tables are at the core of his great astronomical work, the *Almagest*. In addition to its importance for astronomy and (eventually) surveying, trigonometry gave scientists their first real example of a function with a continuously varying input.

Outline

I. Astronomy was one of the dominant forces behind the development of mathematics, and in this lecture we'll see how astronomy led to the development of trigonometry.

- **A.** Most people think of trigonometry in connection with land measurement—and today it is used in surveying—but it was originally applied to the study of astronomical phenomena.
- **B.** We also think of trigonometry as being defined in terms of the ratio of the sides of a right-angle triangle, but again, that is not the way people have thought of the trigonometric functions—the sine, the cosine, and the tangent function—throughout most of the history of mathematics.
- **C.** In this lecture, we will see that those functions emerge out of the problem of determining the length of a chord of a circle, that is, the straight-line distance between two points on a circle.
- **D.** Trigonometry is extremely important because it introduces the function: the idea of a process that yields a well-defined output no matter what real number is input.

II. Let's begin by thinking about the solar system.

A. Most of us have a picture of the solar system in our minds: a bunch of balls, representing the planets, traveling in elliptical paths around a ball in the middle, representing the Sun.

B. Of course, no one has ever actually seen this picture of the solar system. What astronomers see is simply the night sky.

1. The first thing we observe by looking at the night sky is that the relative positions of the stars are fixed. This enables us to identify constellations: certain groupings of stars that always stay in the same relationship.
2. Even though the relative positions of the stars are fixed, the stars themselves move in the sky throughout the night. The one star that does not seem to move is the North Star, or the polar star, but all the others turn in a great celestial sphere around it.
3. Ancient observers noticed the fact that the Sun constantly changes position. They also realized that knowing the position of the Sun against the dome of the night sky is important for knowing when to plant, when to expect rains, when to harvest, and so on.

C. The path of the Sun is called the *ecliptic*, a term that comes from the means that astronomers found for determining the position of the Sun.

1. It is difficult to see where the Sun is against the stars because when the Sun is shining, the stars around it are not visible. Ancient astronomers used lunar eclipses to discern the path of the Sun.
2. When the Sun, the Earth, and the Moon are aligned, the Moon is at the exact opposite point in the sky from the Sun. During a lunar eclipse, the Moon is located where the Sun will be exactly six months in the future.
3. With this realization, astronomers were able to gradually plot out the points where the Sun would be at different times of the year. Collecting lunar eclipses over a long period of time enabled fairly accurate tracking of the Sun.
4. The signs of the zodiac are references to the constellations that give a rough idea of where the Sun is at different times of the year. An illustration in the

frontispiece of Ptolemy's *Almagest* shows a model, called an *armillary sphere*, of the relative positions of the Sun, the various stars, and Earth, along with the signs of the zodiac.

D. Ancient observers also noticed that some stars move. In Greek, these became known as the *planetes*, or "wanderers," what today we call a *planet*. These stars also follow the ecliptic, the same path as the Sun. Astronomers tracked the planets because they believed that their positions might also have an important relationship to events on Earth.

III. Aristotle's view of an Earth-centered universe consisted of a great sphere of the stars, with the Sun and the planets traveling on the path of the ecliptic, but this model ran into some problems.

A. One of the first problems to be observed was that of *retrograde motion*, that is, the fact that the planets do not always move forward along the curve of the ecliptic. At times, they seem to slow down and back up.

B. We understand this phenomenon today because we realize that other planets, Mars, for example, circle the Sun, not the Earth. The Earth also circles the Sun, but because our planet is closer, it takes less time for Earth to travel around the Sun than it does Mars. At times, when Earth comes closest to Mars, we pass Mars, but from our perspective, Mars seems to be backing up. Once the Earth is far enough away, it looks like Mars has begun to move in its normal direction again.

C. The first person to find an explanation for the problem of retrograde motion was Aristarchus of Samos (c. 310–230 B.C.).

1. Aristarchus suggested that retrograde motion could be explained if both Mars and Earth circled the Sun, rather than Mars circling the Earth. His solution was considered preposterous.

2. People realized that if the Earth were circling the Sun, the planet would have to be traveling at incredible speeds. (In fact, the speed of the Earth around the Sun is about 67,000 miles an hour.) If our planet were traveling that fast, they thought we would have to be aware of it.

D. The solution to retrograde motion that was generally adopted came from Apollonius of Perga.

1. Apollonius suggested that rather than Mars circling the Earth, Mars is actually traveling on a smaller circle, called an *epicycle*, whose center circles the Earth.
2. This answer preserved one of the foundational Aristotelian viewpoints about the nature of the heavens—that all the events in the cosmos are based on circles.

IV. Another problem encountered in observation of the heavens is that the seasons are not all the same length. This fact was observed by Aristotle and explained by Hipparchus of Rhodes (190–120 B.C.).

A. The ancient astronomers knew the location of the Sun at both the summer and the winter solstice; these two points are exactly opposite each other on the great circle of the ecliptic. Drawing right angles from the line that connects the summer solstice to the winter solstice shows the vernal, or spring, equinox and the autumnal equinox. The seasons are marked by the time between these points.

B. If the Sun were traveling in a circle with the Earth at the center, each of the seasons should be the same length, but they are not. Winter, from the winter solstice to the spring equinox, is about 89 days long. Summer, from the summer solstice to the autumnal equinox, is more than 4 days longer.

C. Hipparchus of Rhodes saw that this difference could be explained if the position of the Earth was slightly off center in the universe. If the Earth is slightly off center and is closer to where the Sun is in winter, then the Sun actually travels a shorter arc in winter and a longer arc in summer.

D. Hipparchus's attempts to figure out the degree to which Earth was off center led to the problem of calculating the length of a chord, for which Hipparchus invented trigonometry.

1. Imagine that we have an arc of a circle and we want to figure out the length of the straight line that connects the two points at the ends of that arc.

2. First, we have to decide how we will measure the arc of a circle; we make this measurement in degrees, an idea that was inherited from the Babylonians.
   a. Today, we use degrees (°) to measure how much something has turned. If something has turned 90°, that means it has made a quarter-turn. This usage, however, is fairly recent.
   b. Up until the 18th century, degrees were a measure of arc length, which is a distance around the outer edge of a circle. The Babylonians divided the circle into 360°, almost certainly because that is approximately the number of days it takes the Earth to travel around the Sun.
   c. These degrees of arc length were then subdivided, according to the Babylonian system, into $\frac{1}{60}$ of a degree, which we call a *minute* (′). The minute is divided into $\frac{1}{60}$ of $\frac{1}{60}$ of a degree, or a *second* (″).
3. Given any particular arc length, we want to find the length of the chord that connects the endpoints of the arc.
4. That measurement will depend on the value of the radius. As we make the radius larger, the chord length changes; thus, the chord length is always described in terms of the radius.
5. If we have an arc of 90° (one-quarter of a circle) and we know the radius, the chord of 90° of arc is the radius multiplied by $\sqrt{2}$.
6. If we have a chord that corresponds to an arc of 60°, an equilateral triangle is formed by the two radial lines that go out to the circumference; thus, the chord of 60° is exactly 1 radius.
7. The Greeks were also able to figure out the exact value of a chord of 72°, which is $\sqrt{\frac{5-\sqrt{5}}{2}}$ times the radius.

E. The trigonometric functions come from these chords.
   1. Think about an arc and the chord that connects the two endpoints of an arc. Using the terminology of bows and arrows, the arc is the bow and the chord is the string. The line from the center of the circle to the midpoint of

the arc is the arrow. If we connect the radial lines that go from the center of the circle to the bottom of the arc and the top of the arc, the result looks like a bow and arrow.

2. If we consider not the full chord length but only half of the chord length (from the point where the chord attaches to the arc at the top down to the arrow), we are looking at the sine.
3. The Greeks did not work with the sine; as we'll see in the next lecture, this idea as well as the cosine came out of India. The tangent function was invented by Islamic astronomers.

V. The greatest astronomer and one of the greatest mathematicians of the Hellenistic period was Ptolemy of Alexandria (c. A.D. 100–c. 170), a scientist who shared the name of some of the rulers of Egypt.

A. Ptolemy put the ideas of Hipparchus into an incredible work on astronomy that runs to 13 books. His work explained astronomical phenomena, just as Euclid's *Elements* had explained mathematical principles. This great work of Ptolemy's was originally known as the *Mathematiki Syntaxis* (*Mathematical Collection*), but it has come down to us today as the *Almagest*.

B. In one part of the book, Ptolemy constructed a table of values that lists arc lengths and corresponding chord lengths.

1. Again, the circumference of a circle is 360°. It is easiest to work with chord lengths if the chord length is measured in the same units that are used to measure arc length.
2. With a circle of circumference 360°, the radius needed is $\frac{360}{2\pi}$, which is a little bit more than 57°. Ptolemy and other astronomers usually converted the circumference to $360 \times 60'$; then, the radius in minutes is approximately 3438.
3. Ptolemy explained how to use the chord lengths for two different arcs to find the chord length of their difference. For example, knowing the chord lengths for 72° and 60°, he could find the chord length for 12°.

4. Ptolemy also explained how to use the chord length for a given arc to find the chord length for half that arc. Thus, if we know the chord length for 12°, we can find the chord length for 6°, 3°, $\frac{3}{2}$°, and so on. This method does not, however, yield the chord length for 1°.
5. One of the advantages of using the same unit of measure for the arc length and the chord length is that as we work with smaller and smaller arcs, the ratio between the chord length and the arc length approaches 1. For very small arcs, we can assume that these two values are approximately equal.
6. This means that an arc of $\frac{3}{4}$° divided by a chord of $\frac{3}{4}$° should about equal an arc of 1° divided by a chord of 1°. This tells us that the chord of 1° is very close to the reciprocal of $\frac{3}{4}$ (which is $\frac{4}{3}$) times the chord of $\frac{3}{4}$.
7. Knowing the exact value of the chord of $\frac{3}{4}$, Ptolemy could then find a close approximation to the chord of 1°, and with that he could find the chord of $\frac{1}{2}$°. He then used his formula for the sum of two arc lengths to construct his table of chord lengths in intervals of $\frac{1}{2}$°. The chord length for $\frac{1}{2}$° corresponds to the sine of $\frac{1}{4}$°.

VI. These astronomical works that came out of the eastern Mediterranean would be imported to India, where astronomers would carry the ideas of Ptolemy much further and develop many new ideas in mathematics.

Suggested Readings:

Heath, *A History of Greek Mathematics*, chap. 17, 245–97.

Katz, *A History of Mathematics*, chap. 4, 135–57.

Questions to Consider:

1. Up through the time of Newton and into the early 18th century, degrees were considered a measurement of the length of an arc of a circle rather than a measurement of angle, though either is easily translated to the other. What are the advantages and

disadvantages of thinking of degrees as a measurement of the arc of a circle?

2. A *radian* is a unit for measuring angles in which 2π represents a full turn of 360°. Why would radians be a natural unit to use if, instead of angles, we measure the length of the arc of a circle?

Lecture Five—Transcript

Astronomy and the Origins of Trigonometry

This is the last of our lectures on Greek mathematics. In this lecture I am finally going to start talking about astronomy. As I said, astronomy has been one of the dominant moving forces behind the development of mathematics, and in this lecture we are going to look at how astronomy led to the development of trigonometry.

Most people think of trigonometry as something that is used for land measurement, and certainly today it is used in surveying, but that is not where we get trigonometry. Trigonometry actually comes to us from astronomy—from problems in astronomy that I am going to talk about in this lecture—and its application to problems of surveying actually didn't come about until quite recently. The earliest actual reference that we have to the use of trigonometry for the purposes of surveying is a book written in A.D. 1595. For most of the history of the development of trigonometry, it came from astronomy, and it was being used exclusively in order to study astronomical phenomena.

We also think of trigonometry as being defined in terms of the ratio of the sides of a right-angle triangle. Again, that is not the way trigonometry has been thought of—that is not the way people have thought of the trigonometric functions (the sine, the cosine, and the tangent function) throughout most of the history of mathematics. We are going to see that those functions come out of the problem of trying to determine the length of a chord of a circle—finding the distance, the straight-line distance, between two points on a circle. This idea of defining the trigonometric functions in terms of ratios of sides of a right-angle triangle only comes about in the 18th century.

Trigonometry is extremely important—not just for the trigonometric functions and the role that they are going to play in mathematics in many different ways (that turns out to be one of those abstractions that is extremely important in many different contexts), but they also begin to really introduce the idea of function. I will talk more about this in this lecture and then again in the next lecture—this idea of having a process where you can take an input that can be any real number whatsoever, and you need to have a well-defined output no matter what input you put into this function.

I need to start by thinking about the solar system, and most of us have a picture of the solar system: There is a standard picture where you see a bunch of balls. There is a bright yellow ball in the center that represents the Sun, and you've got a small blue ball that is the Earth, and you've got a little red ball that is Mars, and you've got all of these balls that are lying out there. You see nice elliptical paths that are marked out, and the planets are going in these nice elliptical paths around the Sun. Some people wonder how the ancients could have been so confused that they didn't realize that the planets go around the Sun.

Of course, this traditional picture of the solar system that we have is something that nobody has ever seen. Nobody has ever actually put a satellite into that position and taken a picture of the solar system—and even if they did, that is not what they would have seen. The planets are much too small to show up as nice round balls—they simply would be dots of light. And planets don't leave vapor trails as they travel around the Sun—you don't get these nice elliptical paths coming out. What the astronomers have seen is simply the night sky, and it is important to begin by thinking about what you can see in the night sky and how the astronomers began this process.

The first thing that you observe by looking at the night sky is that the relative position of the stars is fixed, and this is why we crate the constellations—why we are able to identify constellations, certain groupings of stars that always stay in the same relationship. But even though the relative position of the stars stays fixed, the stars themselves are moving in the sky throughout the night—and, in fact, if you take a time-lapsed photograph pointing toward the north, what you will see is that the stars are actually moving, and you can see the tracks of these stars over time. The one star that stays fixed is the North Star, or the polar star, which does not move, but all of the other stars are turning in a great sphere or great celestial sphere around this North Star.

One of the things that was observed in antiquity far into prehistory—we have no idea when this first was observed—but the fact that the Sun does not have a fixed position against this dome of the stars, but in fact the Sun is constantly changing its position, and you can see that in the fact that the first stars that you see on the horizon at sunset or the last stars that you see on the horizon at sunrise are going to be different constellations at different times of the year.

It was quickly realized that the position of the Sun against the dome of the night sky is very important. Knowing where the Sun is located tells you where you are in the process of the year—so it tells you when to expect the rains and when to expect spring to come, and it tells you when to plant and when you should expect that the harvest is ready to be brought in. The position of the Sun against the heavenly dome is an extremely important thing to be aware of, and the actual path of the Sun is what we call the *ecliptic*—that comes from the means that astronomers found for determining where the Sun is located.

It is very difficult to see where the Sun is against the stars, because when the Sun is out it is so bright that you can't see the stars around it. Ancient astronomers figured out the path of the Sun by using lunar eclipses. When the Sun, the Earth, and the Moon are all lined up, the Moon is exactly at the opposite point of the sky from where the Sun is located—and so the Moon, during a lunar eclipse, is located where the Sun will be exactly six months in the future. By realizing this, then astronomers were able to gradually plot out the points where the Sun would be at different times of the year, and once you begin to collect lunar eclipses, over a period of 700 years you get a very accurate track of where the Sun is. The position of the Sun at different times in the year then tells you what to expect from the seasons.

What we get are the signs of the zodiac, which are references to the constellations, which give a rough idea of where the Sun is located. There is a nice illustration of this in a frontispiece from Ptolemy's great work, the *Almagest*, that I will talk about a little bit later in this lecture, and what you see in this frontispiece is what is called an *armillary sphere*—it is an object that shows the relative position of the Sun and the various stars. The Earth is located in the center of the armillary sphere, and you see the band of the ecliptic, which is the path that the Sun moves along—and you also see the signs of the zodiac on this armillary sphere.

There are a number of problems that come up with the ideal situation for what is happening with the motion of the Sun and what is happening with the stars, and one of the things that was observed fairly early on is that in addition to the fixed stars, there are stars that move. In Greek, they became known as the "wanderers," or *planetes*, from which we get our word "planet"—so the planets are the stars that move, and it was observed that these stars that move also follow the ecliptic. To the ancient astronomers, or astrologers, or mathematicians—again, those three terms really are synonymous—they observed that these points of light followed the same path as the Sun. If the position of the Sun was so important for what is happening on Earth, then the position of these smaller lights must also be important for what is going on on the Earth. They are much smaller than the Sun, and so one would expect their influence to be much more subtle—but nevertheless, they should have an important influence, and this is why astronomers and astrologers started tracking the position of the planets to figure out what is going on.

This was Aristotle's view of an Earth-centered universe with a great sphere of the stars, the Sun, and the planets traveling on this path of the ecliptic, but it ran into some problems. One of the first of these to be observed was the problem of *retrograde motion*—the fact that the planets do not keep moving along this great curve of the ecliptic. Every once in a while, they will slow down and back up. This was a problem that Aristotle put out to the scientists of that time, the mathematicians and astronomers of that time, to explain what is going on. If, in fact, we have circular motion going out in the heavens, why do the planets occasionally stop, back up, then stop again and start going in the original direction?

We understand this phenomenon today because we realize that a planet, such as Mars, is circling the Sun, not the Earth. The Earth is also circling the Sun, and the Earth is closer to the Sun, so it takes less time for the Earth to go around the Sun. What happens is every once in a while the Earth is going to lap Mars, so when the Earth comes closest to Mars, it is actually going to pass it, and it is going to look like Mars is backing up. Then, once the Earth is far enough along, it will again look like Mars has begun moving in its normal direction—in this case, around the Sun—or what it looks like from the Earth, that Mars is going around the Earth.

This problem of trying to explain retrograde motion, this backing-up motion of the planets, was something that Aristotle posed, and the first person to come up with a solution was Aristarchus of Samos. He was born sometime around 310 B.C., and Aristarchus suggested that maybe retrograde motion is coming from the fact that Mars does not go around the Earth; maybe Mars and the Earth both go around the Sun. He actually came up with the correct solution—but, of course, nobody believed him because his solution was totally preposterous. There is no possible way that the Earth could actually be circling the Sun; we would have to be traveling at incredible speeds. (In fact, the speed of the Earth around the Sun is about 67,000 miles an hour.) How could we possibly be traveling around the Sun at that kind of a speed and not be aware of it? People began to look for other solutions.

The person who eventually would come up with the solution to retrograde motion that would generally be adopted was Apollonius of Perga. This is the same Apollonius that we met in the last lecture who was famous for his work on conics, and he is the one who suggested that what is really going on is that as Mars circles the Earth, Mars itself is actually located on a little circle whose center circles the Earth. What happens is that we have Mars, which is going around and around a little circle, and the center of this little circle is going around the Earth. What is actually happening is that Mars executes a kind of looping motion as it travels around this little circle called an *epicycle*, so an outer circle, and it goes around this little circle (the epicycle) as the center of the epicycle goes around the Earth. This preserved one of the foundational Aristotelian viewpoints about the nature of the heavens, which was that everything that goes on in the heavens must be built out of circles.

There are other problems with trying to explain what goes on in the heavens, and another one of the problems that actually was observed by Aristotle and finally solved by Hipparchus of Rhodes (who lived sometime in the 2nd century B.C.) was the problem that if you look at the length of the seasons, they are not of equal length. The ancient astronomers knew where the Sun was located at the summer solstice, and they knew where the Sun was located at the winter solstice, and these two points are exactly opposite along this great circle of the ecliptic. You then look at right angles from the line that connects the summer solstice to the winter solstice, and that gives you the vernal equinox (or the spring equinox) and the autumnal equinox. The seasons are marked by the time from the winter solstice to the spring equinox, and that gives you winter; the spring equinox to summer solstice, which gives you spring; the summer solstice to the autumnal equinox, which gives you summer; and so on. One would expect if the Sun were just traveling in a nice big circle with the Earth at the center that each of the seasons would be the same length, but they are not.

It may not seem this way, but actually winter is the shortest season, and summer is the longest, and the difference between the two is not subtle. Winter is about 89 days long—from the winter solstice to the spring equinox. Summer is more than 4 days longer. It is 93 and about 5/8 of a day—so it is more than 4 days longer from the summer solstice to the autumnal equinox, and it was Hipparchus of Rhodes who came up with the idea of how to explain this.

It could be explained if, rather than putting the Earth at the center of the universe, you had the Earth slightly off center. So, if the Earth is slightly off center, if it is closer to where the Sun would be in winter, then you are going to get a smaller arc that the Sun has to travel during winter, and you are going to get a longer arc that the Sun has to travel during summer. Of course, this immediately begins to upset the Aristotelian worldview, because it moves the Earth off the center of the universe, but that seems to be the way to do it. Hipparchus was looking at this problem and trying to figure out how off center the Earth had to be. That is where we get to the problem of calculating the length of a chord.

In order to do this, Hipparchus essentially invented trigonometry. What he did was to invent a way of determining the length of the chord that connects two points on the arc of a circle. This would become the basic problem of trigonometry. You've got an arc of a circle, and you want to figure out the length of the straight line that connects the two points at the end of that arc.

First of all, you've got to decide how you are going to measure that arc of circle, and we measure it in degrees—and this has been inherited from the Babylonians, who measured the arc of a circle in degrees, and this is something else that people very often do not understand. Today, we use degrees to measure how much something has turned. We say that something has turned 90°—that means it has made a quarter-turn. That is very recent. That is an 18th-century interpretation of degrees. Up until the 18th century, degrees were a measure of arc length—they were a measure of how far you had traveled around the outer edge of a circle. The Babylonians divided the circle into 360°, almost certainly because that is approximately the number of days that it takes the Sun to travel completely around the Earth—or the Earth to travel around the Sun.

Interestingly, the Chinese divided a circle into 365¼°. This is nice in one respect because it means that the Sun moves exactly 1° every day, but working with a circle of 365¼° is very difficult to do, and so the Babylonians made a compromise: They picked an easier number to work with—360. What that means, though, is that the Sun does not move exactly 1° across the ecliptic every day. It moves a little bit less than 1° every day.

These degrees of arc length, then, would be subdivided according to the Babylonian system into sixtieths of a degree, which we call a *minute* [′]. The word "minute" actually comes from Latin—it means "a small part," *pars minuta*, and so we take the word *minuta*, and it became, in English, "minute." Then, you divide the minute up into 60 seconds of arc, so 1/60 of 1/60 of a degree—and this is called, in English today, a *second* [″], and this comes from the phrase: "the second small part" (*pars minuta secunda*)—and so *secunda* (the second small part) became our "second."

We are measuring the arc length in degrees, minutes, and seconds. Given any particular arc length, we want to find the length of the chord that connects the two endpoints of that arc. That is going to depend on the value of the radius—so as I make the radius larger, the chord length is going to change—and so the chord length is always described in terms of the radius. If I take an arc of 90°, that is a quarter of a circle, and if I know the radius, it is simply the length of a diagonal, and so the chord of 90° of arc is going to be the radius multiplied by $\sqrt{2}$.

If I take a chord that corresponds to an arc of 60°, I simply get an equilateral triangle formed by the two radial lines that go out to the circumference—and so the chord of 60° is exactly 1 radius. The Greeks were also able to figure out the exact value of a chord of 72°, which turns out to be the square root of the quantity 5 minus the square root of 5, all divided by 2: $\sqrt{\frac{5-\sqrt{5}}{2}}$.

I have said that this problem of finding chord lengths is where we get trigonometry—and in fact trigonometry does correspond to these chords. If we think about taking an arc and then taking the chord that connects the two endpoints of the arc (and the terminology here is the terminology of bows and arrows—so an arc or a bow), and you've got the chord (or string) that connects the two endpoints of that arc, and we consider the line from the center of the circle out to the midpoint of our arc, which was referred to as the arrow (it looks very much like an arrow), we can now connect the radial lines that go from the center of the circle down to the bottom of the arc and the center of the circle up to the top of the arc. It really does look like a bow and arrow at this point.

If you consider not the full chord length but only half of the chord length—so you consider the chord that goes from where it attaches to the arc at the top down to the arrow, what you have there is precisely the sine. The Greeks did not work with the sine. The Greeks worked with what today we would think of as twice the sine, the chord. This idea of working with half of that quantity—working with the sine—is something that would come out of India, and I will talk about that in the next lecture, how Indian astronomers picked up the ideas of Hipparchus and the other Greek astronomers and then developed them into the kind of trigonometry that we understand today. Instead of working with a full chord, they worked with a half-chord (what they call the sine), and then you also get the other trigonometric functions. The tangent itself would be invented by Islamic astronomers.

In the time that remains I want to talk about the greatest astronomer of the Hellenistic period and also one of the greatest mathematicians of this period, and that is Ptolemy of Alexandria. We have run across a number of Ptolemys before: Ptolemy I, II, and III, who were rulers in Alexandria, rulers of Egypt. This is a scientist who happened to share the same name. He lived during the 2nd century A.D., and he took the ideas of Hipparchus, and he put them into an incredible work on astronomy—a work that runs to 13 books, a work that is so comprehensive and does such a great job of explaining astronomical phenomena that just as when Euclid's *Elements* came out people threw away any mathematical works they had from before Euclid—the same is true with Ptolemy.

When Ptolemy's astronomical work came out, everybody threw away any other astronomical texts that were out there, and so none of the texts from before Ptolemy actually exist. All we have are a few references to what was contained in them, and it is Ptolemy's own references to Hipparchus that tell us what Hipparchus had been doing.

This great work of Ptolemy was known as the *Mathematiki Syntaxis* (the *Mathematical Collection*), and it would come to be known as the *Great Collection* (the *Megisti Syntaxis*)—the Islamic astronomers would then translate this into Arabic, and they would come to know it simply as "The Great Work" (*al-Magisti*), taken from the *Magisti Syntaxis*. When this then was translated from the Arabic into Latin for European astronomers and European scientists, they took the Arabic term, the *al-Magisti*, and they called the work the *Almagest*—and so that is the term by which it is commonly known today.

One of the great results that Ptolemy was able to do in this book—there is a lot of astronomy in it, and there is a lot of mathematics in it—but I want to focus on one part, which is Ptolemy's efforts to construct a table of values where you could put in the arc length that you were interested in and then read off what the chord length would need to be. As I said, he started with certain chord lengths that were fairly easy to determine, and of course, one of the things that you've got to decide as you come into this is what your radius is going to be. The circumference of a circle is 360°. It is going to be easiest to work with chord lengths if we measure the chord length in the same units that we use to measure arc length. We really want to measure the chord length in degrees.

If I've got a circle of circumference 360, that means that I am going to need a radius that is 360 divided by 2π, which is a little bit more than 57°. It is about 57°17′45″. That is very difficult to work with—so what Ptolemy and the other early astronomers usually did was to convert everything to minutes, so the total circumference of the circle would be 360×60, and then the radius in minutes is approximately 3438.

What Ptolemy was able to do, then, was to take the values of chord length that he knew. He knew the chord length for 72°, and he knew the chord length for 60°, and he also was able to work out how the chord length of a given arc could be determined if you knew the chord lengths of two arcs that added up to that. Or equivalently, if you know the chord length of one arc and you know the chord length for a smaller arc, you have found the formula for finding the chord length of the difference. This is equivalent to our modern formulas for finding the sine of a sum of angles, or the sine of a difference of angles, and so using these he was able to find an exact value for the chord length of 12°, and then you can also use this formula if you know the chord length for a given arc, you can find the chord length for half of that arc. So, once you know the chord length for 12°, you can find the chord length for 6°, or 3°, or (3/2)°, or (3/4)°, but that doesn't give you the chord length for 1°—and this is something that Ptolemy really wanted to work out to quite a bit of accuracy. One of the advantages of using the same unit of measure to measure the arc length and the chord length is that as we take smaller and smaller arcs, the ratio between the chord length and the arc length should approach 1.

For very, very small arcs, we can assume that they are approximately equal. That means that the (3/4)° arc divided by the chord of (3/4)° should about equal the arc of 1° divided by the chord of 1°. That tells me that the chord of 1° is very close to the reciprocal of 3/4, which is 4/3, times the chord of 3/4. He knew the exact value of the chord of 3/4, and he could then use that in order to find a very, very close approximation to the chord of 1°.

Once he had the chord of 1°, he could find the chord of (1/2)°, and once he had the chord of (1/2)°—he had this formula: If you know the chord of one arc length and the chord of another arc length, you can add them together. So he was able to build up an entire table of chord lengths in intervals of (1/2)°. And if you know the chord length for (1/2)°, that corresponds to the sine of (1/4)°.

Effectively, what Ptolemy was able to construct was a table of values of the sine down to just (1/4)° at a time. Ptolemy was 2nd century A.D. (as I have said), and Greek mathematics really ends about A.D. 400. Fortunately though, this was not the end of the developments in astronomy, because these astronomical works that came out of the eastern Mediterranean would be imported into India. The Indian astronomers would pick up this idea and carry it much further and also bring in many new ideas in mathematics—and, in particular, one of the things that we will see in the next lecture is how this work on trigonometry would lead to a general appreciation for polynomials.

Lecture Six

Indian Mathematics—Trigonometry Blossoms

Scope:

This lecture surveys early Indian mathematics, much of which is difficult to reconstruct because it was recorded as brief mnemonic descriptions written in verse in such texts as the Vedas and Sulbasutras. Indian methods for calculating and recording numbers were similar to those used in China. The great period of Indian mathematics occurred during the Gupta Empire, when astronomers with access to Alexandrian astronomical texts made significant advances in trigonometry and the related problem of finding formulas that would enable accurate interpolation of trigonometric tables. The great astronomical center of Ujjain would be the heart of Indian mathematics until the 13th century. This lecture concludes with a brief coda on the development of infinite series in Kerala during the 14th and 15th centuries.

Outline

I. The origins of Indian mathematics are rather uncertain.

   A. The earliest records come to us from the Vedas, the epic poems that were written in the middle to late 2nd millennium B.C. The mathematics contained in the Vedas is sketchy and open to interpretation; often, we find only mnemonic devices for mathematics in these verses.

   B. A collection of works that are appendices to the Vedas, known as the *Sulbasutras* (written c. 800–200 B.C.), describe how rituals were to be performed, lay out plans for constructing altars, and so on. These texts contain more mathematics than the Vedas.

      1. The Sulbasutras explain how to find approximations to $\sqrt{2}$ and $\sqrt[3]{2}$ to enable priests to double the area and volume of an altar. Doubling the area requires multiplying the length and width by $\sqrt{2}$, and doubling the volume requires multiplying each of the three dimensions by $\sqrt[3]{2}$.

2. These poetic texts also contain references to the Pythagorean theorem, as well as interesting counting problems. One of the ways in which the texts were memorized was to recombine the syllables in the lines in different ways. Some of the earliest work in the field of mathematics called *combinatorics* comes from this endeavor.

C. The ancient Indian people used a system for recording numbers that is much closer to the system we use today.

1. The Indians had nine symbols for the digits 1 through 9, which could be combined with special symbols for the powers of 10. Under this system, to represent the number 327, for example, we would write the digit 3, the 100s symbol; the digit 2, the 10s symbol; and the digit 7.
2. This was not yet a full place-value system in the modern sense. The system had no digit for zero, which we use as a placeholder to mark the difference between, for example, 307 and 37.
3. As we will see in the next lecture, this system was also used by the Chinese, but scholars do not know where the system originated.

D. Zero was invented sometime between A.D. 300 and 600/700.

1. The earliest record we have of zero being used as a placeholder in a number is in a Hindu temple in Cambodia constructed in the year A.D. 683. As recorded in the Hindu system, the year of construction was a number that used the digit 0.
2. Zero had been used a bit earlier, not as a placeholder but as a number that could be manipulated, by the 7th-century astronomer Brahmagupta. He explained how to add or subtract zero and how to multiply by zero.

II. This lecture focuses on three great Indian astronomers: Aryabhata (476–550); Brahmagupta (598–c. 665); and Bhaskara Acharya, or Bhaskara the Scholar (1114–1185).

A. As mentioned in the last lecture, the astronomical works developed in Alexandria and elsewhere in the eastern Mediterranean would be taken up by Indian astronomers in the Kushan Empire.

1. The Kushan Empire had connections with an older Hellenistic empire, the Seleucids, who were based in Persia and central Asia.
2. Ptolemy's *Almagest* does not seem to have been among the astronomical texts brought to northern India during the Kushan Empire. Many other works of the same period, however, were brought to India and translated into Sanskrit. Indian astronomers built on these texts, further developing the ideas of trigonometry.

B. The first of the great Indian astronomical texts is the *Surya Siddhanta* (written c. 4th or 5th century A.D.). Even in this early work, we see that the Greek chord has been replaced by the half-chord. The half-chord would become known in the West as the *sine*, a word that originates from an Arabic mistranslation of a transliteration of a Sanskrit word.

C. Indian astronomers also studied the cosine, called the *kotijya*.
1. Recall again the picture of the bow and arrow from the last lecture; the shaft of the arrow—the distance from the center of the circle to the chord—is the cosine.
2. The tip of the arrow between the chord and the arc of the bow, called by Indian astronomers the *ukramajya*, would become in English the *versed sine*. We no longer work with the versed sine, because it is simply the radius minus the cosine.

III. We turn now to the work of Aryabhata with trigonometric tables.

A. Aryabhata lived during the period of the Gupta Empire (A.D. 320–600) in Kusumapura, a center for astronomical work near the capital of Pataliputra, which today is modern Patna in India.

B. Aryabhata also worked with a circle of radius 3438, although he almost certainly was not aware of Ptolemy's *Almagest*. That text had found chord lengths for arcs down to $\frac{1}{2}°$, but Aryabhata worked with chord lengths in multiples of 3°45′.
1. In other words, he constructed his table by starting with a chord length for 60° and then found the chord lengths for 30°, 15°, $7\frac{1}{2}°$, and $3\frac{3}{4}°$ by taking half-angles.

2. Aryabhata's was a much rougher table than Ptolemy's, but he did important work with the problem of interpolation: How do we go about finding intermediate values within a table of values?
3. If we have a value that is halfway between two given values, a natural approach to interpolation is to take the output that is halfway between the two known outputs. Below is a table that will help us see how this idea works.

Input	Output
1	1
2	3
3	7
4	13

4. If our input is $2\frac{1}{2}$, what should the output be? The fact that $2\frac{1}{2}$ is halfway between 2 and 3 suggests that the output should be halfway between 3 and 7. The logical choice for the output is 5. This is called *linear interpolation*, which assumes that the inputs and outputs work as if they are on a straight line.
5. However, if we look at the output values, we see that the differences between them are increasing. The difference between 1 and 3 is 2; the difference between 3 and 7 is 4; and the difference between 7 and 13 is 6.
6. As we move further down the table, the differences increase; thus, the increase in the output going from 2 to $2\frac{1}{2}$ should not be as big as the increase in output from $2\frac{1}{2}$ to 3.
7. The following table agrees with the original table for integer inputs and gives us outputs at the half-integers so that the difference between the outputs increases by the same amount.

Input	Output
1	1
$1\frac{1}{2}$	$1\frac{3}{4}$
2	3
$2\frac{1}{2}$	$4\frac{3}{4}$
3	7
$3\frac{1}{2}$	$9\frac{3}{4}$
4	13

8. We can now see that a better output for an input of $2\frac{1}{2}$ is 3 plus an increase of $\frac{7}{4}$, or $4\frac{3}{4}$.
9. This approach is called *quadratic interpolation*. The Indian mathematician Brahmagupta showed how to do an arbitrary quadratic interpolation for any kind of input function.

IV. Brahmagupta was an extremely important mathematician and astronomer. He taught at the great astronomical center of Ujjain, which is located on the western edge of Madhya Pradesh, near modern Rajasthan.

A. Ujjain was founded in the 4th century B.C. and would continue until the 12th century. It was so important as a mathematical and astronomical center that it would mark zero longitude, or the first meridian, for Indian astronomers. The first meridian in our modern world is the line of Greenwich, chosen because of the royal astronomical observatory there.

B. As mentioned earlier, Brahmagupta was the first person known to have used zero and one of the first to use negative numbers. He realized the need to work with negative numbers and accept negative numbers as possible solutions.

C. The quadratic interpolation formula is often attributed to Newton, who extended it to interpolation by polynomials of higher degree. But Indian astronomers discovered this idea first.

D. Brahmagupta also did work on Diophantine equations, such as the Pythagorean triples we saw earlier ($3^2 + 4^2 = 5^2$).

1. Brahmagupta studied such Diophantine equations as $x^2 - 8y^2 = 1$. The ratio of any two integers that satisfy this equation gives a good approximation to $\sqrt{8}$.
2. Brahmagupta found a number of solutions to that equation: $3^2 - 8(1)^2 = 1$, $17^2 - 8(6)^2 = 1$, and $3363^2 - 8(1189)^2 = 1$.
3. Another example of a Diophantine equation studied by Brahmagupta is $x^2 - 61y^2 = 1$. One of his solutions was: $1{,}766{,}319{,}049^2 - 61(226{,}153{,}980)^2 = 1$.

V. Brahmagupta's ideas would further be developed by another great Indian astronomer, Bhaskara Acharya (1114–1185).

A. Bhaskara was interested in the problem of finding polynomials that interpolate values in tables of sines, cosines, and versed sines. The approach he came up with anticipated developments in western Europe that would come many hundreds of years later.

B. To find a polynomial to approximate the sine, Bhaskara considered the rate at which the sine function changes. In looking at the derivative of the sine function, he realized that the rate at which the sine function changes is given by the cosine function, and the rate at which the cosine function changes is given by the negative of the sine function.

C. Bhaskara saw how to use this fact to find what today we would call a *Taylor polynomial*, a quadratic polynomial that produces a good approximation to the sine function. If we know the exact values of the sine and cosine, we can work out this quadratic approximation.

D. Bhaskara's work marked an increase in the understanding of polynomials; up until his time, quadratic polynomials had been thought of as useful only for solving problems that involve areas. Cubic polynomials were thought of in terms of volumes.

1. Now that polynomials were used to approximate missing values in a table, there was no reason to stop at third-degree polynomials.
2. The Greeks stopped with third-degree polynomials because it made no sense to talk about a

four-dimensional region, but with interpolation it makes sense to consider polynomials of degrees four, five, six, and so on.

3. It seems that this was one of the important sources of the development of the idea of a polynomial of general degree.

E. The type of Diophantine equation studied by Brahmagupta (e.g., $x^2 - 8y^2 = 1$) is known today as *Pell's equation.* Bhaskara found a procedure for solving Pell's equation no matter what multiplier is used for the second perfect square, as long as the multiplier itself is not a perfect square.

VI. Indian mathematics would continue after the Delhi caliphate conquered Ujjain in 1235, although the astronomical center was destroyed and the astronomers in Ujjain were dispersed.

A. The mathematical work of the time moved into southwestern India, in Kerala, where we find such astronomers as Madhava in the 14th and early 15th centuries, Paramesvara in the late 14th century and into the 15th, and Nilakantha in the 15th century and into the early 16th.

B. These astronomer-mathematicians furthered the general idea of Bhaskara of using the rate of change of a trigonometric function—and the rate of change of the rate of change—to find a quadratic polynomial that is a good approximation to the sine function, followed by a cubic polynomial, a fourth-degree polynomial, and eventually, a polynomial of arbitrary degree.

C. Madhava is generally credited with figuring out how to find polynomials of infinite degree, what today we call a *power series*. This is a representation of the function as a sum of powers of the unknown, with the powers going out to infinity.

D. Unfortunately, the astronomers who were working on power series in southern India did not communicate their results to anyone else. Because no other uses for the sine or cosine were known at this time, knowledge of the power series was lost. It was only in the 19th century, when British archaeologists found some of these Indian texts, that mathematicians realized what these earlier thinkers had accomplished.

VII. In the next lecture, we'll move north from India to China to see what was happening there at the same time.

Suggested Readings:

Bressoud, "Was Calculus Invented in India?"

Datta and Singh, "Hindu Trigonometry."

Katz, *A History of Mathematics*, chap. 6, 210–32.

Varadarajan, *Algebra in Ancient and Modern Times*, 17–31.

Questions to Consider:

1. Today, trigonometric functions are usually introduced as ratios of sides of right triangles. Which do you consider to be conceptually simpler: the half-chord of a circle or the ratio of the opposite side to the hypotenuse in a right triangle?
2. By 1500, the most advanced mathematics in India—the infinite summation—was the preserve of a small group of devotees who explored it purely for its own sake. It died stillborn. Are there any lessons in this for us today?

Lecture Six—Transcript
Indian Mathematics—Trigonometry Blossoms

In the last lecture we finished our study of Greek mathematics, and in this lecture we move on to Indian mathematics.

The origins of Indian mathematics are rather uncertain. The earliest records that we have actually come to us from the Vedas, the great epic poems that were written in the middle to late 2nd millennium B.C. The mathematics that is contained in the Vedas is very sketchy, and often what is found there is more of a mnemonic device, and so it is open to a great deal of interpretation what exactly is being described in the Vedas. There is a collection of works that really are appendices to the Vedas—what are known as the *Sulbasutras*—written in the period roughly 800 B.C. to 200 B.C. These describe how rituals were to be performed, and they lay out the way in which altars were to be constructed, things such as this, and so there is a fair amount of mathematics that is contained in the Sulbasutras. One of the problems that the Indian priests would face is one of doubling the area of an altar, and in order to double the area, what you need to do is take the length and width and multiply each of those by $\sqrt{2}$, and so the Sulbasutras explain how to find a good approximation to $\sqrt{2}$. They also explain how to compute volumes: If you want to double the volume of a given altar, you need to multiply each of the three dimensions by $\sqrt[3]{2}$, and so these also explain how to find good approximations to $\sqrt[3]{2}$.

We do find references to the Pythagorean theorem here. We also find references to interesting counting problems, because one of the ways in which the Vedas, and the Sulbasutras, and these other poetic forms were memorized was to recombine the syllables. People were often interested in the number of ways that these could be recombined, and some of our earliest work on what we call *combinatorics* actually comes from these kinds of investigations.

The earliest recorded writing that we have of Indian mathematics also makes it clear that they were using a very different system for recording numbers, a system that is much closer to the system that we use today. They had nine symbols for the digits 1 through 9, and then they would combine these with special symbols for the powers of 10. So you had a symbol for 10, a symbol for 100, a symbol for

1000, and so on, and you put these together. So if you wanted to represent a number like 327, you would write the digit 3, the 100s symbol, the digit 2, the 10s symbol, and then the digit 7. It was not yet a full place-value system in our modern sense, and one of the problems was that at this point, the Indian mathematicians did not have the digit 0—they did not have this placeholder that enables us to tell the difference between 307 and 37—and so they used symbols for the power of 10 that was being recorded.

This system, as we will see in the next lecture, was also used by Chinese mathematicians. One of the interesting questions out there is whether this system was originally created in India and then imported into China, or originally created in China and then imported into India, or perhaps created someplace else—such as central Asia—and then imported into both of those countries, or maybe it just happened to come up independently. It was an idea that made good sense in both India and China, but at roughly the same time we see a very similar system of writing the numbers.

So, zero itself would be invented sometime—nobody is really sure—sometime between A.D. 300 and about A.D. 600 or 700. The earliest actual recording that we have of zero being used as a placeholder in a number is actually in a Hindu temple in Cambodia that was constructed in the year A.D. 683, and the year as recorded in the Hindu system was a number that actually used the digit 0.

So, zero had been used a bit earlier—not as a placeholder, but as a number that could be manipulated, by the 7th-century astronomer Brahmagupta. This really is the first time that we have record of somebody working with a symbol that represents nothing, a symbol for zero. So, zero now enters the numerical system, and Brahmagupta did explain what happens when you add a zero to a number, or subtract zero from a number, or what happens if you multiply a number by zero—you still get zero. He wrestled a bit with this whole issue of what happens when you divide a number by zero—and essentially, he sort of left that up in the air, what happens when you try to divide by zero.

In today's lecture I am going to focus on three great Indian astronomers. Again, these were men who were best known for their contributions to astronomy, but they also did a great deal of work in mathematics generally—in algebra, in number theory, and other areas of mathematics. These are Aryabhata, born about A.D. 476 and died sometime in the middle of the 6th century A.D.; Brahmagupta, whom I have already mentioned, who lived in the 6th century; and then Bhaskara Acharya—Acharya is an honorific title that means "scholar," so this is really "Bhaskara the Scholar"—who lived in the middle of the 12th century.

As I said in the last lecture, the astronomical works that were developed in Alexandria and elsewhere in the eastern Mediterranean would be taken up by the Indian astronomers, and this appears to have happened in the first several centuries A.D. under the Kushan Empire. The Kushan Empire was an empire that had come down out of central Asia to take over northern India. It had strong connections with an older Hellenistic empire, the Seleucids, who were based in Persia and also in central Asia, and largely because of this the Kushan Empire maintained strong connections with the eastern Mediterranean. We do know that they brought astronomical texts from Alexandria and elsewhere.

Strangely, it seems that they never actually imported Ptolemy's *Almagest*, and that appears to be one of the books that was not translated into Sanskrit, but many of the other astronomical works of that same period were brought in to India and were translated into the Sanskrit. Fairly soon after receiving the Greek texts, the Indian astronomers began to build on those and further develop their idea of trigonometry.

The first of the great Indian astronomical texts is the *Surya Siddhanta*. We don't know exactly when it was written—probably in the 4th century or 5th century A.D. It is possible it was written even earlier than that, but already here in this first Indian astronomical work that uses trigonometry we see that the Greek chord has been replaced by the half-chord—what would come in the West eventually to be called the *sine*.

There actually is an interesting story behind how we got this word "sine." The Indian astronomers quite naturally called it a "half-chord." The word for "chord" is *jya*, and "half" is *ardha*—so this was *ardha-jya*, or "half-chord," although eventually they dropped the adjective "half," and they just referred to it as a *jya* or *jiva*. This meant though the "half-chord," and from *jiva* the word was taken up by the Islamic astronomers, and they pronounced it as *jiba*, and they wrote it in Arabic—which only uses consonants—as *jb*. When the first Western astronomers were trying to figure out what this function was that was being described, they looked at these Arabic letters *jb*—which, of course, did not represent any word in Arabic. *Jiba* is not an Arabic word; it is simply a transliteration of the Sanskrit word for this "half-chord." They looked at those letters *jb*, and the one Arabic word that they knew that used the letters *jb* was *jaib*, which means "breast," and so they assumed that this was to be called a "breast." It is not quite clear how this happened, but perhaps out of excessive modesty, rather than calling it a breast, they decided to refer to the fold of a cloth that might cover a breast, and the fold of a cloth could be referred to as a *sinus*, and so this became a *sinus* in Latin, which in English eventually would become our "sine"—so we get "sine" out of a mistranslation of a transliteration of a Sanskrit word.

The Indian astronomers also studied the cosine, which was the *kotijya*, and if you think about the bow and arrow, so we've got the arc of the bow, we've got the chord, and we've got the arrow, the shaft of the arrow—the distance from the center of the circle to the chord—that is the cosine or the *kotijya*. The Indian astronomers also made great use of the *ukramajya*, and that is the tip of the arrow, which is the little part of the arrow between the chord and the arc of the bow. This would come to be called in English the *versed sine*. We no longer work with the versed sine. The versed sine, after all, is just the radius minus the cosine, so there is no real need to come up with a separate term for this, but if you look in the 17th century, in something like Newton's *Principia*, his *Mathematical Principles of Natural Philosophy*, he does a lot with the versed sine, and this is where it is coming from. It is an Indian invention, and it comes out of describing that arc, that tip of the arrow.

I would like to spend some time now looking at the trigonometric tables that people came up with, and particularly the work of Aryabhata. Aryabhata, as I said, was born in 476. He was in the Gupta Empire. The Gupta Empire had begun around A.D. 320 and would last until about A.D. 600. Aryabhata lived in Kusumapura, which was the center for astronomical work, very near the capital of Pataliputra, which today is modern Patna in India. He also was working with a circle of radius 3438, and he almost certainly was not aware of Ptolemy's *Almagest*, because as I said, Ptolemy's *Almagest* was actually able to look at chord lengths for arcs all the way down to (1/2)°. Aryabhata was working with chord lengths of 3°45′. In other words, what he was doing was starting with a chord length for 60°, and then from that you can get the chord length for 30, 15, 7½, and then 3¾ just by taking half-angles. So he had a much rougher table, but he did something that was extremely important. He was looking at the problem of interpolation: If you've got a table of values, and you want to find some intermediate value, how do you go about doing that?

A natural way of doing this interpolation is that if you've got a value that is halfway between the two given values, you just take the output that is halfway between the two outputs. And rather than looking at the actual trigonometric table that Aryabhata was looking at, I am going to show you a somewhat simplified table that illustrates the point of what Aryabhata and the later Indian astronomers were able to do.

What I am going to do, on one side I have got inputs that are the digits 1, 2, 3, and 4, and then I have got outputs that are the digits 1, 3, 7, and 13. The question then is: If I've got an input of 2½, what should be my output? Well, 2½ is halfway between 2 and 3, so that suggests that my output should be halfway between 3 and 7—so we might take 5 as the logical output, and that is what is called a *linear interpolation*. We are just assuming that things work as if they are on a straight line. But in fact, if you look at these output values—1, 3, 7, and 13—you see that the output values are increasing. So, 1 to 3 is a difference of 2, and 3 to 7 is a difference of 4, and 7 to 13 is a difference of 6. These differences are themselves increasing.

Aryabhata did not just look at the differences, he actually took a table of the second differences, and he used that to refine the way in which he was going to do this interpolation. What is happening in this table is that as we move further down the table, the increase gets bigger, and so the increase in the output going from 2 to 2½ should not be as big as the increase in output as we go from 2½ down to 3. We need a bigger increase in output. The total increase in output from 2 to 3 has got to be 4—so the output increases from an output of 3 to an output of 7, but how do we then arrange this output in order to take into account the fact that the outputs are increasing at an increasing rate? One of the things that you can do is to decide that from 1 to 1½, we've got to have a certain increase in the output;

from 1½ to 2, we've got to have a greater increase in the output; and the increase in the output has to increase by the same amount each time. So we can do this if from 1 to 1½ we have an increase of 3/4; from 1½ to 2 we have an increase of 5/4; from 2 to 2½ we have an increase of 7/4; and then from 2½ to 3 we have an increase of 9/4. Now our increases are going up by the same amount each time: 7/4 plus 9/4 gives us our 16/4, and our total increase of 4 as we go from 2 to 3, so that says that a better output for an input of 2½ is 3 plus an increase of 7/4, so that gives us 4¾ as the desired output.

What is happening here is that you are beginning to get the idea of taking this table of values and doing what today is called a *quadratic interpolation*, so you are no longer using just a straight line to figure out what number should come out. You are using a much more sophisticated kind of technique for interpolation, and this idea of using the quadratic interpolation would be more fully developed by Brahmagupta.

Brahmagupta would actually show how to do an arbitrary quadratic interpolation for any kind of function that you might put in. So if you know the value of first differences and you know the values of the second differences, you can then use those in order to get a better idea of what the intermediate values in your tables should be.

Brahmagupta is an extremely important mathematician and astronomer. He taught at the great astronomical center of Ujjain, which is located on the western edge of Madhya Pradesh, near modern Rajasthan. Ujjain was founded around the 4th century B.C. and would continue until around the 12th century A.D. It would be an important mathematical center, and it would be so important as an astronomical center that it would mark zero longitude, or what was the first meridian, for Indian astronomers. The first meridian in our modern world is the line of Greenwich, and that is chosen because of the royal astronomical observatory in Greenwich. It is decided that that is zero longitude, and you measure the longitudes east and west from Greenwich. In ancient Indian astronomy, that zero mark of longitude was located going right through the center of Ujjain, the great astronomical center.

As I mentioned before, Brahmagupta is the first person known to have used zero. He also is one of the first people known to have used negative numbers. We will see in China that negative numbers came in much earlier, but only as intermediate steps. Brahmagupta realized that you needed to be able to actually work with negative numbers and accept negative numbers as possible solutions—and, so far as we know, he is the first person actually to accept negative numbers as solutions to mathematical problems.

This interpolation formula, this quadratic interpolation, is often attributed to Newton, who would then go on and extend this to interpolation by polynomials of higher degree, cubic polynomials, fourth-degree polynomials, and so on. Eventually, the Indian astronomers would realize how to do this, how to not just take first differences and second differences, but to work with third differences, fourth differences, and as many as you needed in order to get a very accurate interpolation of the values.

Brahmagupta also did work on Diophantine equations. These are equations that were studied by Diophantus, and one of the examples of these that I gave earlier is the problem of Pythagorean triples, so: (3, 4, 5); ($3^2 + 4^2 = 5^2$). The kind of Diophantine equations that Brahmagupta studied, equations where we are only interested in integer solutions, are things like $x^2 - 8y^2 = 1$. If you find two integers that solve this, then the ratio of those integers gives you a very good approximation to $\sqrt{8}$, a good rational approximation. He found a number of solutions to this, and we don't know exactly the method that he used. All we know are the answers that he found—for example, he showed that $3^2 - 8(1)^2 = 1$, and $17^2 - 8(6)^2 = 1$, and $3363^2 - 8(1189)^2 = 1$.

Another one of the examples of the Diophantine equations that Brahmagupta studied are something squared minus 61 times something else squared is equal to 1 ($x^2 - 61y^2 = 1$). Here the answer gets much more complicated. One of the answers that Brahmagupta actually came up with for this is you take 1,766,319,049 and you square it, and you subtract 61 times 226,153,980—and this quantity gets squared—and then that difference is actually equal to 1 [$1{,}766{,}319{,}049^2 - 61(226{,}153{,}980)^2 = 1$].

This idea would further be developed by the last of the great Indian astronomers that I want to talk about: Bhaskara Acharya, born in 1114 and died in 1185. He also was interested in this problem of finding polynomials that interpolate values in these tables of sines, and cosines, and versed sines. He came up with something that would really anticipate developments in western Europe that would come many, many hundreds of years later—and that is in order to find a good polynomial to approximate the sine, he considered the rate at which the sine function is changing. What Bhaskara actually was looking at is the derivative of the sine function, and what he realized is that the rate at which the sine function changes is given by the cosine function, and the rate at which the cosine function changes is given by the negative of the sine function. He saw how to use this fact in order to find what today we would call a *Taylor polynomial*—a quadratic polynomial that does a very good approximation to the sine function, very close to any value that you happen to know the exact value of the sine function for. If you know the exact value of the sine, you know the exact value of the cosine, and you can work out this quadratic approximation. This really marks a new episode in the understanding of polynomials, because up until now polynomials had been thought of simply as things that are useful for solving problems that involve areas, and then you use a quadratic polynomial or problems involving volumes, and then you are going to use a cubic polynomial, but now suddenly polynomials are being used for interpolation. They are being used to approximate the missing values in a table, and once you start doing that, there is no reason to stop at polynomials of degree three. There is no reason to stop with volumes.

The Greeks stopped with polynomials of degree three because it made no sense to talk about a four-dimensional region, but if you are simply looking at the interpolation, suddenly it makes sense to go beyond degree three and consider a polynomial of degree four, a polynomial of degree five, a polynomial of degree six. We are not entirely sure where these different polynomials came from, but it seems that this was one of the important sources of the development of the idea of a polynomial of general degree.

Also, the Diophantine equation that was studied by Brahmagupta, this general problem, you've got a perfect square and you subtract some integer times another perfect square—I gave the example where it is a perfect square minus 8 times another perfect square equals 1 ($x^2 - 8y^2 = 1$), or a perfect square minus 61 times another perfect square is equal to 1 ($x^2 - 61y = 1$), and that is known today as *Pell's equation*, and this is an equation that was not just studied by Bhaskara. Bhaskara actually was able to find a way of coming up with a solution to Pell's equation no matter what multiplier you were using. You could use an 8, or you could use a 61, or you could use any number as long as it is not a perfect square. You could take $x^2 - 57y^2 = 1$ or $x^2 - 327y^2 = 1$, and Bhaskara laid out a procedure that would enable you to find a pair of integers that would then satisfy that equation.

Indian mathematics would continue after the fall of Ujjain. Ujjain disappears in 1235. This is when the Delhi caliphate conquers Ujjain, and the astronomical center is destroyed at that time, at the time of the conquest, and the astronomers in Ujjain are dispersed. Northern India ceases to be a center of important mathematical work, but the mathematical work of that time does move down into southern India, specifically southwestern India in Kerala, and we do have a series of important Indian astronomers doing work in mathematics down in Kerala—beginning with Madhava in the 14th century into the early 15th century; Paramesvara, late 14th century into the 15th century; and Nilakantha, 15th century into the early 16th century.

What these astronomer-mathematicians were able to do is to take the general idea that Bhaskara had done, of taking the trigonometric function, looking at its rate of change, and using the rate of change, and the rate of change of the rate of change, and the rate of change of the rate of change of the rate of change to come up first as Bhaskara did with a quadratic polynomial that is a good approximation to the sine function, and then a cubic polynomial, and then a fourth-degree polynomial—and eventually, they saw how to get a polynomial of arbitrary degree.

So you can get a polynomial of as high a degree as you want; the higher you take the degree, the closer the approximation is going to be. It is not clear which of these astronomers actually was responsible for it. It is generally believed that this goes back all the way to Madhava, but at this time what they did was not just figure out how to get polynomials of arbitrarily large degree, they then made the jump to a polynomial of infinite degree—what today we would call a *power series*. This is an infinite collection of powers of your unknown that can be brought as close as you wish to the actual value of the function—so it is a representation of the function as a sum of powers of the unknown. The powers go out toward infinity, and you can get as close to the value as you want just by going far enough out.

This is work that was probably done in the 14th century, and this is work that would not be redone in Europe until the very end of the 17th century. Unfortunately, the work that was done in India was lost to the rest of the world. The astronomers who were working on power series in south India were not able to communicate their results to anyone else, and the fact is that their way of approximating the sine was almost too accurate. The problem was that you can only measure the difference between two lights in the sky to within maybe a couple minutes of arc—so being able to calculate values of the sine, or the cosine, or the tangent to this kind of accuracy was far more than what was needed.

At this point, nobody could figure out any other uses for the sine or the cosine; the only uses for these trigonometric functions at this time was in astronomy, and since no one could figure out these uses, eventually, knowledge of these power series would be lost. It was only in the 19th century, actually when British archaeologists managed to come across some of these old documents from several hundred years earlier, that mathematicians looking at them realized what actually had been accomplished. It points out the importance of not just pulling ideas from the world around you; the ideas that you pull out of the world around you have got to be applicable to worthwhile problems in understanding the world. Once the mathematics becomes divorced from what is really needed, there is the danger that it is going to degenerate.

In the next lecture I am going to move north from India into China, and we will see what was happening at this same time, the 1st millennium A.D., roughly up to about A.D. 1300 in China.

Lecture Seven

Chinese Mathematics—Advances in Computation

Scope:

The earliest surviving Chinese mathematical writing is from the Western Han dynasty. It contains indications of what was known in the 3rd century B.C. and suggests a long mathematical tradition in China. Chinese mathematics was of a practical nature, and by the 1st century A.D., mathematicians there were using advanced computational techniques. The Han mathematicians used a base-10 number system with decimal fractions and employed negative numbers. By the 5th century A.D., Chinese mathematicians had found an approximation to π that would be the most accurate known for another 900 years. During the period A.D. 1000–1200, the Chinese did sophisticated algebra that included interpolation techniques and summation formulas that would not be rediscovered in Europe until the 17th and 18th centuries.

Outline

I. Thus far in the course, we've seen that mathematical ideas spring from different sources, including the fields of surveying and astronomy. Chinese mathematics grew primarily out of the needs of civil administration.

A. The record of mathematics in China begins with the early part of the Han dynasty (also known as the Western Han dynasty, or former Han dynasty, 206 B.C.–A.D. 25). This was the first Confucian dynasty, in which people were trained in mathematics, philosophy, and jurisprudence to become civil administrators.

B. Chinese textbooks consist primarily of problem sets, similar to the textbooks we saw in Babylon and Egypt.

1. Of the two important manuscripts we have from this time, the first is the *Zhou bi suan jing* (*Mathematical Classic of the Zhou Gnomon*, written c. 1st century B.C.). This text contains problems of similar triangles related to surveying. The work also states the Pythagorean theorem and methods for calculating square roots.

2. The other manuscript is the *Jiuzhang suanshu*, or *Computational Prescriptions*. This text deals with finding areas and volumes and includes the Pythagorean theorem, as well as linear equations and rules for calculating square roots and cube roots. The problems in this text involve surveying, commerce, and tax collection.
3. The same system for representing numbers that was used in India is used in the *Computational Prescriptions*, along with decimal fractions. The Chinese extended the idea of the units place, tens place, hundreds place, and thousands place in the other direction, looking at tenths, hundredths, and thousandths.
4. At about the same time, the 2nd or 3rd century B.C., negative numbers were also used as intermediate steps in some calculations.

C. Buddhism was introduced in China around the 1st century A.D., and shortly after that, we begin to find Indian astronomical texts in China. The Chinese began to work with trigonometric functions based on what they had learned from India. It would not be until about the year A.D. 1000, however, that the Chinese would begin to use zero as a placeholder.

II. The first Chinese mathematician we know by name is Liu Hui (fl. late 3rd century A.D.), who is best known for his work on surveying, recorded in the *Haidao suanjing*, or the *Sea Island Computational Canon*.

A. The title of the work comes from its first problem, which involves determining the height of a mountain on an island by an observer who is somewhere offshore and does not know his distance from the base of the mountain. Liu Hui explains how to use the idea of similar triangles to work out the height of the mountain.

B. Liu Hui's work also contains the first Chinese proof of the Pythagorean theorem. In addition, he explains a general method for finding areas and volumes that is equivalent to the Greek method of exhaustion.

C. Liu Hui used the same method for approximating π as Archimedes.

1. Both first found the circumference of a regular hexagon inscribed inside a circle and then doubled the number of sides.
2. Archimedes stopped when he got to a regular polygon of 96 sides, but Liu Hui continued to a regular polygon of 192 sides. He actually states that π is approximately 3.14, giving the tenths and hundredths in the decimal.
3. Liu Hui further stated that decimals could be expanded to as many places as needed, which may have been a new idea at the time. An approximation to the value of any distance could be found by using progressively smaller units.

D. The most accurate approximation to π from this time comes from Zu Chongzhi (fl. late 5th century A.D.). His approximation, $\frac{355}{113}$, is accurate to a degree of error of less than 3 parts in 10 million. He, too, may have found this approximation using polygons with progressively more sides.

III. Another problem the Chinese worked with during this period involves applying numbers to time.

A. When we try to apply numbers to time, we are working with different units—the day, the lunar month, and the solar year. How do we decide when these units line up?

B. Obviously, astronomers in Egypt and Babylon must have worked with this problem, but the first record we have of a solution for aligning the different cycles comes from a Chinese work called *Sunzi suanjing*, or *Sunzi's Computational Canon* (written c. A.D. 280–473). This solution today is called the *Chinese remainder theorem*.

1. The example given in the *Sunzi suanjing* is as follows: Consider three cycles of lengths 3, 5, and 7. Knowing that we are 2 units into the cycle of length 3, 3 units into the cycle of length 5, and 2 units into the cycle of length 7, how long ago were all these cycles lined up?
2. The situation is equivalent to finding a number that will have a remainder of 2 when divided by 3, a remainder of 3 when divided by 5, and a remainder of 2 when divided by 7.
3. Sunzi does not explain how he found the answer, 23.

IV. The next significant mathematician to come along was Li Chunfeng (A.D. 602–670), director of the astronomical observation service and chief astronomer and astrologer of the Tang dynasty (A.D. 618–907).

A. Li Chunfeng pulled together, corrected, and explained all the mathematical work from various texts that had been written in China up to the period of the Tang dynasty.

B. His work is known as *The Ten Computational Canons* (written A.D. 644–648).

V. Jia Xian, a court eunuch, was a Chinese mathematician from slightly after the year 1000 (fl. c. mid-11th century).

A. Jia Xian is generally credited with being the first person to come across Pascal's triangle, a triangular arrangement of numbers with 1s along each side and such that the numbers inside the triangle are obtained by adding the two numbers that are diagonally above them.

```
                1
            1       1
        1       2       1
    1       3       3       1
1       4       6       4       1
```

B. Pascal's triangle is named for Blaise Pascal because of his work with it in the 17th century. It arises from the problem of expanding binomials.

- We begin with the binomial $1 + x$.
- Multiply it by itself: $(1 + x)(1 + x) = 1 + 2x + x^2$.
- Multiply that result by $1 + x$: $1 + 3x + 3x^2 + x^3$.
- Multiply that result by $1 + x$: $1 + 4x + 6x^2 + 4x^3 + x^4$.

The coefficients are exactly the numbers in Pascal's triangle, giving us an easy way to find the coefficients as we take progressively larger powers of $1 + x$.

C. The reason Jia Xian studied these binomial expansions is that they offer a powerful way of approximating the value at which a polynomial is equal to zero, also known as the *root* of the polynomial.

D. For some polynomials, this value is an exact integer or, perhaps, a fraction. Most of the time, however, this value is an irrational number; thus, the best solution is a good decimal approximation to the number in question, which can be found using Pascal's triangle.

E. Jia Xian's ideas were further developed by Li Zhi (1192–1279), who wrote a work known as the *Ceyuan haijing*, meaning *Mirror Like the Ocean, Reflecting the Heaven of Calculations of Circles* (written 1248). This book contains many geometric techniques, but it also addresses the problem of finding roots of polynomials of arbitrarily high degree.

VI. The mathematician Qin Jiushao (c. 1202–1261) was the first person to explain the Chinese remainder theorem and to show how to use it in any situation with any number of cycles of any length.

A. One of the examples Qin Jiushao gives is as follows:

1. Find a number that is 32 units into a cycle of length 83, 70 units into a cycle of 110, and 30 units into a cycle of 135.
2. In other words, find the smallest positive number that gives a remainder of 32 when divided by 83, a remainder of 70 when divided by 110, and a remainder of 30 when divided by 135.

B. The number is 24,600.

VII. The culmination of Chinese mathematics came in the late 13th century, under the rule of Kublai Khan and slightly thereafter.

A. One of the great works from this time was the *Siyuan yujian*, the *Trustworthy Mirror of the Four Unknowns*, written by Zhu Shijie (c. 1260–1320) around 1303.

1. As mentioned in an earlier lecture, Isaac Newton is credited with discovering a general formula, known as the *Newton interpolation formula*, for interpolating polynomials using first differences, second differences, third differences, and so on.
2. However, the *Siyuan yujian* shows that Zhu Shijie and other Chinese mathematicians at the end of the 13th

century clearly knew the Newton interpolation formula in its full generality.

B. Zhu Shijie also worked with Pascal's triangle to find the sum of the binomial coefficients along a diagonal. His result was later known as *Vandermonde's formula*, named after an 18th-century European mathematician.

C. Chinese mathematics began to disappear after the 13th century, probably because of the chaos that arose in China after the fall of Kublai Khan.

VIII. In the next lecture, we'll turn to Islamic mathematics and the emergence of algebra. We know that Islamic mathematics built heavily on work done in India, but how much of it drew on what was happening in China?

Suggested Readings:

Katz, *A History of Mathematics*, chap. 6, 192–210.

Martzloff, *A History of Chinese Mathematics*.

Needham, *Science and Civilisation in China*, 1–168.

Straffin, "Liu Hui and the First Golden Age of Chinese Mathematics."

Swetz, "The Evolution of Mathematics in Ancient China."

Questions to Consider:

1. Our earliest records for Chinese mathematics coincide with the creation of a sophisticated bureaucracy. Why would this favor the development of calculational techniques over the emphasis on logical reasoning that marked Greek and Hellenic mathematics?
2. Even though the Chinese were the first to use negative numbers in their calculations, they considered them to be computational conveniences rather than legitimate numbers. This would also be true of Islamic and European mathematicians well into the 17th century. What are the conceptual difficulties associated with negative numbers?

Lecture Seven—Transcript

Chinese Mathematics—Advances in Computation

In this lecture we are going to look at Chinese mathematics, taking it up roughly to about A.D. 1300.

In my very first lecture I talked about different sources of mathematical ideas—one of which is civil administration, another is navigation and surveying, and the third is astronomy. As I said, all of these come into play in all of the great mathematical cultures, but different ones come to the fore in different cultures. For example, for the Babylonians and the Egyptians, as we saw, the dominant mathematical theme was what was needed for civil administration. When we get to the Greeks, the dominant theme is navigation and surveying, and this would influence the Greek decision to put their emphasis on proportion and on geometry. As we saw in the last lecture with Indian mathematics, really the dominant source of mathematical work came from astronomy. It was really the astronomers who were making the progress in mathematics.

As we now shift to China, we are going to see that we go back to a mathematics that primarily comes out of civil administration, very similar to the sources for the mathematics in Babylon and also in Egypt—although also here navigation, surveying, and astronomy will be very important sources for the mathematics that is created.

Our recorded record of mathematics in China begins with the early part of the Han dynasty (what is known as the Western Han dynasty or former Han dynasty), going from 206 B.C. to A.D. 25. What made this particular time particularly propitious for the development of mathematics connected with civil administration is that this is the first Confucian dynasty in Chinese history. This is a dynasty that is really built around the idea of Confucian scholars, people who were drawn out of the general population and trained in the kind of mathematics, and philosophy, and jurisprudence that they would need in order to be civil administrators.

An important role in the Western Han dynasty was that of the teacher, that of the person who would train the future Confucian scholars, and these people needed textbooks. The Chinese textbooks, in many respects, are similar to the kinds of textbooks that we saw in Babylon and in Egypt. These are problem books. These are books that set out lots of problems that a future scholar would need to be able to solve in various contexts, and then would work out the solutions.

We have two manuscripts from this time. The first of these is the *Zhou bi suan jing*, roughly translated as the *Mathematical Classic of the Zhou Gnomon*. This was written possibly in the 1st century B.C., and it might go back even earlier than that. This deals with problems of similar triangles very much involved with questions of surveying. You find the Pythagorean theorem stated quite clearly in this particular work, and there are also methods for calculating square roots.

Another book that comes out of this same period is the *Jiuzhang suanshu*, or *Computational Prescriptions*, in nine chapters. This does a lot of work with finding areas and volumes—and again, we find the Pythagorean theorem, and we also find linear equations. These now are systems of equations, situations in which you've got several things that are unknown, but you've got several relationships connecting them. Problems like this can be found in the ancient Babylonian mathematics, but we also see it reappearing here in the very earliest-known Chinese mathematics. There are also general rules for calculating square roots and cube roots. The kinds of problems that are dealt with in this particular book involve problems of surveying, problems of commerce, and problems of tax collection. They use the same system for representing numbers that was used in India. The largest number that is actually represented in this particular book is 1,644,866,347,500. They clearly were comfortable with working with extremely large numbers.

Even more important than that, the Chinese at this time—this is still about the 3rd century or 2nd century B.C.—were not only using fractions, but they also were using decimal fractions. They extended this idea of a decimal representation: not just units place, tens, hundreds, thousands, but also moving in the other direction and looking at tenths, hundredths, thousandths. In this book, we do have decimal fractions expressed out to thousandths.

Also—here again, we are talking 2nd century or 3rd century B.C.—the Chinese are using negative numbers. As I said, it would be Brahmagupta who was the first person to really accept negative numbers as legitimate solutions to mathematical problems. The Chinese were not there yet, but they were the first people to use negative numbers as intermediate steps in some sort of a calculation that might need to be done.

Buddhism is introduced generally into China around the 1st century A.D., and some time shortly after that we begin to get the Indian astronomical texts being brought up—texts written in Sanskrit and introduced into China. There are references to Chinese monks who would make a pilgrimage to India and be asked to come back with the mathematical texts, and so there was some transfer of mathematical knowledge from India into China in this time. We know that the Chinese built their ideas of trigonometry and began to work with trigonometric functions based on what they had learned from the Indians—although this is one of the few places where we actually are aware of how mathematics traveled from China to other places, or from other places to China. There is very little other direct evidence of influence between China and the other civilizations that surrounded it. It would take a while for the Chinese to begin to use zero as a placeholder. They did not start using it until many hundreds of years after the Indians had begun to use it, but around the year A.D. 1000, the Chinese were using zero as a placeholder—although well after that, many of the Chinese mathematicians would continue to use special symbols for 10, 100, 1000, or tenths, hundredths, or thousandths.

The first actual Chinese mathematician that we know by name was Liu Hui, and he lived sometime late in the 3rd century A.D., and he is best known for his work the *Haidao suanjing* (or the *Sea Island Computational Canon*). This was really a work that was looking at questions of surveying. The name of this particular work, *Sea Island Computational Canon*, comes from the very first problem that is explained in the book, and that is trying to determine the height of a mountain on an island. You are not on that island—you are somewhere offshore, but you see that island, and you want to determine the height of the mountain. You cannot actually determine your distance from where you are to the base of the mountain, so what you have to do is take several different observations from where you are on this opposite shore. What Liu Hui does is to explain how to use the idea of similar triangles in order to work out what the height of this mountain is going to be. We also very significantly find in this particular work the very first Chinese proof of the Pythagorean theorem.

This is not the first known proof of the Pythagorean theorem—as I said, this was known earlier, probably sometime in Greece, between the time of Pythagoras and the time of Euclid—but Liu Hui does work out actual proof of the Pythagorean theorem. Very significantly, he also works out a general method for finding areas and volumes that is equivalent to the method that was developed by Eudoxus of Cnidus, and then described by Euclid, and used so effectively by Archimedes—this idea of method of exhaustion, and that is used by Liu Hui. Liu Hui also comes up with exactly the same idea for approximating pi [π] that Archimedes had discovered—this idea of first finding the circumference of a regular hexagon that is inscribed inside a circle and then doubling the number of sides.

As I said, Archimedes stopped when he got to a regular polygon of 96 sides. Liu Hui took this a step further. He went all the way up to a regular polygon of 192 sides and used that to come up with the approximation, in which he actually states that π is approximately 3.14—so he gives it as the decimal, using tenths and using hundredths. Liu Hui also does something else. Even though decimal fractions had been used earlier, it is not quite clear how much mathematicians of that time were aware that you could continue decimal expansions as far as you wish. Liu Hui specifically states that decimals can be expanded to as many places as you might wish. If you've got any distance, you can always try to find an approximation to its value by taking progressively smaller units—so taking tenths of tenths of tenths, as far out as might be needed.

I said that Liu Hui came up with a very accurate approximation to π: 3.14. As most Chinese schoolchildren know, the most accurate approximation to π from this time comes from a mathematician who lived late in the 5th century, Zu Chongzhi, and he came up with a rational approximation to π of 355/113. It is a remarkable accomplishment, and we don't know exactly how he achieved it, but he might have been using the same idea that Liu Hui did—of taking polygons with progressively more sides. His approximation, 355/113, is accurate to within an error of less than 3 parts in 10 million, and this would be the most accurate approximation to π that would be found for well over a thousand years.

It is also an easy number to remember, especially if you think of fractions the way the Chinese thought of fractions, and the way they still do—which is that you give the denominator first and then state the numerator. I have always thought this makes a lot of sense, because the denominator tells you what kind of thing you are talking about in the fraction, and the numerator then tells you how many of these you've got. So if you take this fraction, 355/113, and instead describe it as "denominator of 113 and numerator of 355," you've got 113 355, and that is a very easy sequence of numbers to remember. As I said, it gives an extremely accurate approximation to π.

There is another problem that the Chinese worked with at this time, and it is a problem that I have alluded to in thinking about how you apply numbers to time. One of the problems is you try to apply numbers to time, and you've got different units to work with. You've got the day, and you've got the lunar month, and you've got the solar year, and how do you decide when all of these are going to line up?

Obviously astronomers all the way back to the Egyptians and the Babylonians must have worked with this, but the first time we actually see a method of solving this problem, knowing when different cycles are going to line up, comes to us from the Chinese mathematicians. The first time that we see this is in a work called *Sunzi suanjing* (or *Sunzi's Computational Canon*). We don't know exactly when it was written, sometime between A.D. 280 and A.D. 473, but what it does is provides a method for finding out when different cycles are going to line up. It is something that today we call the *Chinese remainder theorem*. The actual example that is given in the *Sunzi suanjing* is the following: We consider a cycle of length 3, and we are 2 units into that cycle, and we consider another cycle of length 5, and we know that we are 3 units into that cycle, and we consider a third cycle of length 7, and we know that we are 2 units into that cycle. Another way of thinking about this and the problem is: How long ago were all of these cycles lined up?

The situation is exactly equivalent to we want to find a number so that when we divide it by 3 we've got 2 left over. How long has it been going through cycles of 3 until we get 2 into a cycle of 3? Take this number, divide it by 5, and we want a remainder of 3. So again, how many times do you go through a cycle of 5 and then go 3 more into it (and we want a number that also divided by 7 leaves you with a remainder of 2)?

The answer to this is 23, but you have to back up 23 time units to get all of these lined up: 23 divided by 3 leaves 2, 23 divided by 5 leaves 3, and 23 divided by 7 leaves 2. In this work, the method for finding these solutions is not explained, but it does give an indication that this was being studied. A little bit later I will talk about one of the Chinese mathematicians who actually did find a method for proving it.

The next significant mathematician to come along comes at the very beginning of the Tang dynasty. The Tang dynasty began in A.D. 618, and it ended in A.D. 907. One of the very important people at the beginning of the Tang dynasty was Li Chunfeng. He was the director of the astronomical observation service—and, of course, as the chief astronomer of the Tang dynasty, he would also be the chief astrologer of the Tang dynasty. If you are setting up a new dynasty and you are deciding where to locate your capital and how to build your main buildings, you need to know what the astrological prognostications are. What is a good time at which to launch a particular endeavor?

Li Chunfeng became very important for his astronomical work, but even more important than that, he looked back at all of the mathematics that had been done in China up until that point, and he was given the task of trying to pull it all together. What had happened over the centuries is that many of these mathematical texts had been corrupted. They had been copied, and recopied, and recopied, and mistakes had crept in. There were places where different versions contradicted each other, and there were different methods in different books that contradicted each other—so Li Chunfeng took on the responsibility of taking all of the mathematics that was known in China at that time and pulling it together in what would come to be known as *The Ten Computational Canons*.

This is a work that was done over the period A.D. 644 to 648, and while it is called *The Ten Computational Canons,* in fact they did pull on more than just 10 mathematical works. There were probably at least 12 or 13 different mathematical works that were combined into this—not only did they take these works and correct the corruptions that had come into them, but they also provided extensive commentaries. Much in the same way that Euclid in 300 B.C. took all of Greek mathematics and reestablished it, clarified it, and made a foundation on which further mathematicians could build—Li Chunfeng did exactly the same thing for Chinese mathematics in the middle of the 7th century.

From here, Chinese mathematicians would begin to work on problems of general polynomials, and it is not quite clear when they began to work with polynomials and exactly how the work that they did with polynomials related to the work that was being done on polynomials in India. In the next lecture I will also be looking at Islamic work in mathematics, and we are going to see polynomials coming in and playing a very important role there.

So we've got three great centers toward the end of the 1st millennium A.D. that are all working at getting to this idea of a polynomial of general degree. We would love to know how they interacted, if there was any kind of exchange. The fact that all three centers of civilization were working on similar kinds of problems at exactly the same time suggests that probably there was some kind of communication, although it probably was quite imperfect.

One of the important Chinese mathematicians from slightly after the year 1000 was Jia Xian. Jia Xian was a court eunuch. He lived sometime in the middle of the 11th century. Jia Xian is credited generally with being the first person to come across Pascal's triangle. Pascal's triangle is a triangular arrangement of integers that has 1s down the outside edge, and then the numbers inside are obtained by adding the numbers that are diagonally above it. So I've got 1, and then two 1s, and the next row is 1, 2, 1. The next row is 1, 3, 3, 1. The next row is 1, 4, 6, 4, 1.

This is named for Blaise Pascal because of the work that he did on this particular triangular arrangement of numbers in the 17th century. In fact, even Blaise Pascal realized that it was much older than himself. His name was associated with it because of all the work that he would do on it.

The place that this comes from is the problem of expanding binomials. I am looking at a binomial, $1 + x$. So it is a two-term algebraic expression, and I want to begin to multiply it by itself: $1 + x$ times itself is $1 + 2x + x^2$. If I multiply that by $1 + x$, I get $1 + 3x + 3x^2 + x^3$. If I multiply that by itself, I get $1 + 4x + 6x^2 + 4x^3 + x^4$. If you look at those coefficients, those are exactly the numbers in Pascal's triangle, and so what Pascal's triangle does is it gives you an easy way of remembering, or finding, the coefficients as you take $1 + x$ and take progressive powers of it.

The purpose of doing this, the reason that Jia Xian studied this particular triangular arrangement—the reason that he was looking at these binomial expansions, ways of getting the expanded polynomial out of $1 + x$ raised to a power—is that this is a very powerful way of approximating the *root* of a polynomial—the place where a polynomial actually is equal to zero. We know that by the 11th century in China, these Chinese astronomers and mathematicians were interested in working with general polynomials and trying to find out those values at which the polynomial would equal zero.

For some polynomials, this happens at an exact integer or maybe a nice fraction. Most of the time the location of this zero is going to be some irrational number, some number that cannot be represented as a ratio of integers; so, the best you can do is to get a good decimal approximation to the number in question. You could use Pascal's triangle—you could use these binomial expansions in order to get these roots to whatever accuracy you actually need.

This would be developed by Li Zhi. Li Zhi lived 1192 to 1279, and he wrote a work that is known as the *Ceyuan haijing*, which means *Mirror Like the Ocean, Reflecting the Heaven of Calculations of Circles*. It is a bit of a flowery title. It is a book that was written in 1248, and it has a lot of mathematics in it. We find many geometric techniques, but he also specifically addresses this idea, the problem of finding roots of polynomials to an arbitrarily high degree. It is Li Zhi who references Jia Xian and says that he got this idea—and, in fact, by the time that Li Zhi is writing, this method of using Pascal's triangle is called "the old method." It was at least a couple of hundred years old at this time. Li Zhi, whose manuscript we do actually have copies of today, lays out in a great deal of detail how to find the roots of an arbitrary polynomial by using Pascal's triangle.

Another one of the Chinese mathematicians who was also working at this time is Qin Jiushao. He lived roughly 1202 to 1261. One of the things that he did was to take the Chinese remainder theorem, the problem of astronomical cycles, and you want to know when was the last time they all lined up. If you know the last time they all lined up, that will enable you to determine the next time they will line up, and the time after that when they all line up.

Qin Jiushao, even though the Chinese remainder theorem had probably been solved earlier, he becomes the first person to actually explain how it works and show how to use it in any situation with any number of cycles of any life. One of the examples that he gives is a number that is 32 units into a cycle of length 83; it is also 70 units into a cycle of 110; and it is 30 units into a cycle of 135. In other words, what we want is the smallest positive number so that if we divide it by 83, the remainder is 32, and if we divide it by 110, the remainder is 70, and if we divide it by 135, the remainder is 30. I will leave that out there as a problem for you to work on. I will give you a hint: The number lies somewhere between 20,000 and 30,000.

Chinese mathematics of this era would really come to its culmination in the late 13th century, under a time when Kublai Khan was the ruler of China. One of the great works that was written at this time is the *Siyuan yujian* (the *Trustworthy Mirror of the Four Unknowns*). Zhu Shijie was the author of this. It actually was written slightly after the time of Kublai Khan. It appeared in 1303, but it was built on a lot of work that had been done in China at the very end of the 13th century. This is an incredible work for the sophistication with which Zhu Shijie was able to work with polynomials.

I had mentioned when I talked about the Indian use of interpolating polynomials that a general formula for interpolating polynomials using not just first differences and second differences, but third differences, fourth differences, and as far out as you might want to go is something that Isaac Newton discovered. It is known as the *Newton interpolation formula*. What we find here is the Newton interpolation formula in its full generality, which clearly was known to the Chinese mathematicians at the end of the 13th century.

Something else that Zhu Shijie was able to do at this time is to work with Pascal's triangle, this triangular arrangement of numbers, and find ways for combining these binomial coefficients. His result would come later to be known as *Vandermonde's formula*, and Vandermonde was a European mathematician of the 18th century. Vandermonde's formula was first discovered by Zhu Shijie. This is a formula for taking the sum of the binomial coefficients that come up on a particular diagonal and then finding what they equal in terms of a binomial coefficient.

The evidence is that Chinese mathematics begins to disappear after the 13th century, and that may well be due to the chaos created in China after the fall of Kublai Khan. China broke up into a lot of small warring states, and mathematics really stops being developed in China for quite a few hundred years.

In the next lecture, we are going to switch to Islamic mathematics and the emergence of algebra, and one of the great questions that I would like you to keep in mind as we come into this next lecture is: How much of the Islamic mathematics actually was drawing on what had been happening in China? We know that Islamic mathematics built heavily on what had been happening in India, but we are not sure how much Chinese mathematics influenced what would happen in the Islamic world.

Lecture Eight

Islamic Mathematics—The Creation of Algebra

Scope:

The development of mathematics in the Islamic world began in Baghdad under the ruler Harun al-Rashid, with the collection and translation of the scientific knowledge that then existed, drawing on the Hellenic, Indian, and Mesopotamian traditions. Algebra began here in the 9th century. The Islamic tradition encompasses many great mathematicians, including al-Kwarizmi, from whose writings we get the word *algebra* and whose name is the origin of our word *algorithm*; ibn Qurra, who made progress on Greek problems in number theory; al-Uqlidisi, whose work shows us the first use outside of China of decimals to represent fractions; al-Karaji, who set the stage for the study of arbitrary polynomials; al-Haytham, who advanced Archmidean techniques for computing areas and volumes; al-Samawal, who brought the study of algebra to a high art; and al-Khayyami, better known as Omar Khayyam, who studied cubic equations.

Outline

I. This lecture explores how the mathematical ideas that emerged in ancient Babylon, Egypt, the eastern Mediterranean, India, and China were drawn together in the Islamic world. Here, we will see the creation of algebra and advances in number theory, among other developments.

A. The Islamic calendar begins in A.D. 622, when the Prophet Mohammed fled Mecca to Medina. We will focus on the Abbasid caliphate, which began around A.D. 750. A few years after the founding of this caliphate, Baghdad became the capital and cultural center of the Islamic world.

B. The Islamic leader who would sponsor much of the mathematical activity in Baghdad was Harun al-Rashid (r. A.D. 786–809). He built a library of scientific works and had them translated into Arabic.

C. Harun al-Rashid's successor, Abu Jafar al-Ma'mun (r. A.D. 813–833), established the Bayt-al-Hikma, or House of Wisdom, a center for scholarly work built around the library.

II. One of the first scholars in the House of Wisdom was Abu Jafar al-Kwarizmi (c. A.D. 790–840).

A. He wrote what is considered to be the first book of algebra in the history of mathematics, entitled *Condensed Book on the Calculation of Restoring and Comparing*.

1. As we've seen, Greek mathematicians worked on problems that we would think of as algebraic, and Diophantus introduced the idea of using a letter as a variable, but algebra itself did not yet exist.
2. Al-Kwarizmi introduced the idea of an algebraic equation with two quantities that involve an unknown, for example: $x^2 = 10x + 22$.
3. Al-Kwarizmi explained how to keep the equation in balance by adding, subtracting, or dividing by the same amount on both sides. This is a process he called "comparing and restoring"; the Arabic word for "restoring," *al-jabr*, is the origin of our word *algebra*.
4. Al-Kwarizmi's name is also the origin of our word *algorithm*, which is used in mathematics to mean a procedure with clearly prescribed steps.

B. Al-Kwarizmi's algebra didn't look like modern algebra. He did not use the kind of algebraic notation that we use today, and he expressed unknowns in words rather than variables. He also did not use the equal sign, which was not invented until the 16th century in western Europe.

C. Al-Kwarizmi recognized the usefulness of the Hindu numeral system, which was a full decimal system that used zero as a placeholder. He helped to spread this representation of numbers across the Islamic world.

D. Most Islamic mathematicians at this time did not use negative numbers.

1. This can be traced in part to al-Kwarizmi, who still thought of the solutions to his algebraic equations geometrically, as the Babylonians and Greeks had. In thinking geometrically, it doesn't make sense to consider, for example, a negative length.
2. The avoidance of negative numbers made solving quadratic equations difficult. With an equation such as $x^2 + 10x = 39$, Islamic mathematicians resisted the idea

of working with −39 in subtracting 39 from both sides and accepted only one of the two solutions to this equation, 3, while rejecting −13.

III. The mathematician Thabit ibn Qurra (A.D. 836–901) also lived in Baghdad and did important work in astronomy, geometry, mechanics, and number theory. Among his interests were *amicable numbers*.

A. If we add the proper divisors of 220—1 + 2 + 4 + 5 + 10 + 11 + 20 + 22 + 44 + 55 + 110—we get 284. If we then add the proper divisors of 284—1 + 2 + 4 + 71 + 142—we get 220. The Greeks called such pairs *amicable numbers*.

B. The Greeks also found another pair of amicable numbers, 1184 and 1210, but it was Thabit ibn Qurra who found the next pair, 17,296 and 18,416.

C. That knowledge eventually would be lost to Western mathematics and rediscovered by Leonhard Euler in the 18th century.

IV. Abu'l Hasan al-Uqlidisi (fl. mid-10th century) lived in Damascus and is best known for his book *Kitab al fusul fi al-hisab al Hindi*, or *The Book of Chapters on Hindu Arithmetic*. This book gives us the first use outside of China of decimals to represent fractions, shown by a hash mark over one of a series of numbers.

V. Abu Bekr al-Karaji (c. 980–c. 1030) was a mathematician living in Baghdad who wrote *al-Fakhri*, which means *The Marvelous*.

A. He explored the ideas of raising an unknown to an arbitrary power or a negative power and representing the reciprocal of the unknown by using a negative power (e.g., $x^{-2} = \left(\frac{1}{x}\right)^2$).

B. Al-Karaji saw that the way exponents combine is useful (e.g., $\left(x^3\right)\left(x^5\right) = x^8$) and stated the general rule for multiplying and dividing with exponents.

C. Al-Samawal, a 12th-century Islamic mathematician, credited al-Karaji with being the first Islamic mathematician to discover Pascal's triangle, but we have no independent evidence of this discovery.

VI. One of the most interesting mathematicians from the Islamic world was Abu Ali al-Haytham, born in Basra, in southern Iraq, sometime around A.D. 965 (d. c. 1040).

A. Al-Haytham was an engineer in Basra, but sometime around the year 1000, he was enticed to Cairo, the capital of the Fatimid caliphate and the site of the al-Azhar mosque, founded in 975 as a center of learning.

1. One of the areas in which he concentrated his efforts was optics, trying to understand how mirrors of different shapes operate.
2. Al-Haytham published 92 scientific works, many dealing with optics, as well as other scientific and mathematical topics; 55 of al-Haytham's books still survive.

B. Al-Haytham worked with the Chinese remainder theorem and came up with an observation that would later be known as *Wilson's theorem.*

1. With a prime number (such as 7), the product of all the integers less than that prime (in this case, the integers 1 to 6) plus 1 will be divisible by the prime ($1\times2\times3\times4\times5\times6=720+1=721\div7=103$).
2. For the prime number 11, the product of the integers 1 through 10 plus 1 is 3,628,801, which is exactly divisible by 11.

C. Al-Haytham also furthered Archimedes's ideas for finding areas and volumes and using the method of exhaustion.

1. Archimedes had solved the problem of finding the volume of a paraboloid. This bullet shape results from rotating a parabola around the line of symmetry.
2. Al-Haytham used half the parabola and rotated it around a vertical line to get a solid that comes to a point, a shape that is a common motif in Islamic mosque art.
3. For al-Haytham to find the volume of his solid of revolution, he had to find a formula for adding up the fourth powers of all the integers from 1 to n.
    a. The formula for finding the sum of the integers from 1 to n had been known since antiquity. The formula for the sum of the squares of the integers from 1 to n, which yields a cubic polynomial in n, was also known to Greek mathematicians.

b. The sum of cubes involves a fourth-degree polynomial, a fact that may have been discovered by Greek mathematicians and was certainly known to Indian mathematicians by the middle of the 1st millennium A.D.

4. Al-Haytham found the fifth-degree polynomial that yields the sum of the fourth powers. He then came up with a general procedure to find the sum of fifth powers, sixth powers, and so on.

VII. Before we close, let's look briefly at two other important mathematicians who appeared in the Islamic world after al-Haytham.

A. The first of these is Ibn Yahya al-Samawal, a Jewish medical doctor who lived in the middle of the 12th century in Baghdad.

1. Al-Samawal understood the full power of polynomials and explained how to divide one polynomial by another, a process today called *synthetic division.*
2. Al-Samawal also worked with Pascal's triangle. He may have been the first person to discover the formula for finding the sum of second powers, third powers, fourth powers, and so on using the numbers in Pascal's triangle.

B. The poet Omar al-Khayyami (1048–1131) was also an accomplished mathematician. He was interested in the problem of finding the exact value of the roots of a cubic polynomial. Although he made progress on this problem, it would not be solved until 400 years later by Italian algebraists.

Suggested Readings:

Katz, *A History of Mathematics*, chap. 7.

———, "Ideas of Calculus in Islam and India."

Van der Waerden. *A History of Algebra*, chap. 1.

Questions to Consider:

1. Early Islamic algebra involved neither letters for the unknown quantities nor equations in the sense that we would represent them today. Instead, Islamic mathematicians described each

"equation" using sentences. Al-Kwarizmi introduced the idea of keeping two expressions in balance as one seeks to isolate the unknown quantity. Why is this considered to be the beginning of algebra?

2. Why was this idea of keeping two expressions in balance such an important conceptual breakthrough?

Lecture Eight—Transcript

Islamic Mathematics—The Creation of Algebra

We have now looked at the development of mathematics in ancient Babylon, in ancient Egypt, and also in the Greek world, in the eastern Mediterranean, in India, and in China. In this lecture we are going to draw all of these different mathematical strands together in the Islamic world, what is usually referred to as "Islamic mathematics." That does not mean specifically mathematics connected with the religion, but rather the mathematics that was encouraged by this entire Islamic culture in the Middle East.

Some of the big ideas that we are going to look at are the creation of algebra, advances in number theory—and eventually, we are going to come back to this Islamic period and look at Islamic art in a much later lecture. The Islamic period, of course, begins with the Prophet Mohammed. The Islamic calendar begins in A.D. 622, when Mohammed flees Mecca to Medina, and I want to focus on the Abbasid caliphate that began around A.D. 750. A few years after the founding of this caliphate, the capital of the Islamic world was moved to Baghdad. For the next several hundred years, Baghdad would be the cultural center of the Islamic world, and it would become a very important center for the development of science in general, and mathematics specifically.

The great Islamic leader who would sponsor a lot of this mathematical activity is Harun al-Rashid, and he ruled from 786 until 809. He decided that he wanted Baghdad to be this center of science and mathematics, and so he began collecting the great works and having them translated into Arabic. We happen to know that the very first two works that he had translated were Euclid's *Elements* and Brahmagupta's astronomical work, the *Brahmasphutasiddhanta*. This shows immediately that Harun al-Rashid was drawing on all of the scientific accomplishments that were in existence, that were accessible at that time—not just from Greece in the Hellenistic world, but also from India and South Asia, and probably also to some extent from China, and then also drawing on the local tradition that there had continued to be a strong astronomical and astrological mathematical tradition in Mesopotamia. Baghdad is located in Mesopotamia, in what today is Iraq, and there still was a strong mathematical tradition that could be drawn on in this particular work.

His successor, Abu Jafar al-Ma'mun, ruled from 813 to 833, and he would go a step beyond this—not just translating works into Arabic and building a great library, but also actually establishing the Bayt-al-Hikma (or House of Wisdom). He set himself to do what Ptolemy I had done in Alexandria many centuries earlier—set up a center for scholarly work that was built around a library where you would bring people together to look at what had been accomplished and create new insights.

One of the first people who did this was Abu Jafar al-Kwarizmi. We don't know the exact dates for al-Kwarizmi. He was probably born late in the 8th century and died sometime in the middle of the 9th century A.D. He wrote one of the most important books in the history of mathematics, the *Condensed Book on the Calculation of Restoring and Comparing*, which sounds like a very strange title. This is actually considered to be the first book of algebra. I have talked a bit about the development of the idea of algebra—how you can go back to ancient Babylon and find problems that today we think of as algebraic problems that were being done by the Babylonians, certainly problems that we would think of as algebraic problems being done by the Greek mathematicians, but algebra itself did not yet exist. I talked about Diophantus and how he introduced the idea of a single letter as a variable, and the idea of an actual notation for a quadratic polynomial or a cubic polynomial, but algebra as it is now considered did not yet exist.

What al-Kwarizmi did was to consider the idea of an algebraic equation, and he is the person who comes up with this idea. The two words that he is using, "restoring and comparing," the actual Arabic words are *al-jabr* and *al-muqabala*, and so this is "restoring and comparing," and he comes up with the idea of setting up an equation where you have two quantities that both involve your unknown. For example, you might have something like the unknown squared is equal to 10 times the unknown plus 22 ($x^2 = 10x + 22$). So, you've got an algebraic expression, and the problem then is to try to find a root or a place where this expression actually is equal, where you've got equality on both sides.

What al-Kwarizmi does is he explains how to keep the equation in balance by adding the same amount to both sides, or subtracting the same amount, or dividing both sides by the same amount. This is the process of "comparing and restoring," and his word *al-jabr* is the origin of our word "algebra"—it is simply a transliteration of the Arabic word.

His name, incidentally—al-Kwarizmi—is also the origin of our word "algorithm," a word used in mathematics to mean a procedure that needs to be followed, and al-Kwarizmi laid out this idea of the procedure needed to be followed in order to keep the equations in balance.

Al-Kwarizmi's algebra didn't look anything like our present algebra. He did not have the kind of algebraic notation that we use today. He did not use single letters to represent the unknowns; he would express those in terms of actual words, talking about the unknown and the square of the unknown. You didn't have an equal sign; that is something that would be invented in western Europe in the 16th century. But you did have this idea of an algebraic equation, and that would be an important breakthrough moment that would really lead to a lot of the development of the idea of algebra.

Al-Kwarizmi is also responsible for taking the idea of Hindu numerals. At this point, the Indians are using a full decimal system in the way that we use it today (they have the 10 digits, 0 through 9), and they are using it in a place-value system, where you don't have to have a special symbol to denote what place you are in. You don't need tens, hundreds, or thousands as symbols because you have the zero as a placeholder. It is al-Kwarizmi who realizes how useful this representation of numbers is, and it is al-Kwarizmi who is really responsible for spreading this idea of how to represent numbers across the Islamic world.

There are some problems that the Islamic algebraists and mathematicians still had to wrestle with, and one of these was the question of whether or not to use negative numbers. In fact, most of the Islamic mathematicians of this time refused to use negative numbers. In some respects, this goes back to al-Kwarizmi. While he was willing to describe an algebraic equation, he still thought about its solution geometrically, in the same way the Babylonians did or the Greeks did. If you are thinking about this algebraic equation geometrically, it doesn't make sense to talk about a negative length, and so negative numbers generally were avoided.

That made life a lot harder when you were trying to describe how to solve a quadratic equation, an equation that involved a quadratic polynomial, because if you've got something like $x^2 + 10x = 39$, you can try to subtract the 39 from both sides, but you don't want to actually work with a negative 39 (-39)—and more important than that, there are actually two solutions to this quadratic equation. There is 3, and the other solution is -13, and al-Kwarizmi and the Islamic mathematicians of his time would only accept 3 as a legitimate solution to this particular quadratic equation.

The next mathematician that I want to talk about is Thabit ibn Qurra. He also lived in Baghdad, from about A.D. 836 to A.D. 901. He did important work in astronomy, geometry, mechanics, and he also worked in number theory. One of the interesting things that he did that I would like to say a little bit about is his work on amicable numbers.

This is actually a problem that comes down to us from the Greek mathematicians, and what the Greek mathematicians would do is to take a number, look at its proper divisors, so the positive integers less than that number that divide evenly into it, and add those up and see what number you get, and then repeat this process. So, if I start with 12, its proper divisors are 1, 2, 3, 4, and 6, and if I add those up I get 16. If I now take the proper divisors of 16—those are 1, 2, 4, and 8, and if I add those up I get 15. If I now take the proper divisors of 15—they are 1, 3, and 5, and if I add those up I get 9. The proper divisors of 9 are 1 and 3—they add up to 4. The proper divisors of 4 are 1 and 2—they add up to 3, and 1 is the only proper divisor of 3. So, this process started with 12, went up to 16, and eventually comes down to 1. Do we always come down to 1? No. A simple example is the number 6. If I start with 6, and I add up its proper divisors—1, 2, and 3—I get back to 6. Six is what the Greeks called a *perfect number*, because when you add up its proper divisors you get back to the number itself.

Another example of a perfect number is 28. Its proper divisors are 1, 2, 4, 7, and 14. Add those up, and I get 28 again. One of the possibilities as you do this is you might come right back to the number that you started with. Something else can happen. If you start with the number 220 and add up its proper divisors, you get 284. If you take the number 284 and add up its proper divisors, you get 220, and 220 turns into 284, and 284 turns into 220, and these are called—and they were called by the Greeks—*amicable numbers*, friendly numbers.

The Greeks found this pair, 220 and 284, and they also found another pair, 1184 and 1210, and it was Thabit ibn Qurra who went on and found the next pair, 17,296 and 18,416. That knowledge eventually would be lost to Western mathematicians—it would not be known to the people working in Europe centuries later. In the 18th century, the great mathematician Leonhard Euler would refind precisely this pair of amicable numbers that Thabit ibn Qurra had found, and he was very excited and quite pleased that he had been able to find this result that had, in fact, been found many hundreds of years earlier.

The next mathematician I want to talk about is Abu'l Hasan al-Uqlidisi—and again, we don't know exactly the years he lived, but he was sometime in the middle of the 10th century. We know that he lived in Damascus, and one of the books that he is best known for is the *Kitab al fusul fi al-hisab al Hindi* (or *The Book of Chapters on Hindu Arithmetic*). As I said, al-Kwarizmi was important for urging the adoption of the Hindu method of representing numbers. This particular book really shows how effectively this Hindu decimal system can be used in order to do calculations. One of the things that we find in this book by al-Uqlidisi is the first use outside of China of decimals to represent fractions.

There is a page from one of his books that I am showing up now, and if you go down to the 10th line of this page from al-Uqlidisi's book, the stretches of Arabic writing that have a line over them represent numbers. If you go down on this 10th line you will see a number that has a hash mark over one of the digits, and that hash mark is used to mark where the units are. Of course, instead of reading the number from left to right, you read the number from right to left in the Arabic fashion, and so you read it from right to left. When you get to the hash mark, that designates that you are now in the ones place, and the next digit is representing tenths, and the digit after that is now representing hundredths. It is al-Uqlidisi who manages to convince the Islamic mathematicians to begin using this idea of decimal fractions.

I would love to know whether they borrowed this idea from the Chinese, who had been using it for many hundreds of years up until now, or if this was an independent discovery of this powerful way of representing fractions.

The next mathematician I want to talk about is Abu Bekr al-Karaji. Again, we don't have exact dates on him, but he was born sometime late in the 10th century and died sometime in the early 11th century, perhaps around 1030. He also would live in Baghdad. The book that he is best known for is a*l-Fakhri*, which means *The Marvelous*, and one of the things that he does that will be extremely important as we look at the further development of mathematics is this idea of taking the unknown and raising it to an arbitrary power. You can take the unknown, and you can square it, or cube it, or take it to the 37th power, and you could also take it to a power that is a negative number. He talked about how to represent the reciprocal of the unknown by using a negative power—x^{-2} is the same as $1/x^2$—so he was working with positive exponents and negative exponents, and he pointed out how useful these exponents are because of the way that they combine. If I have x raised to the third power, and I multiply that by x raised to the fifth power, I can combine those just by adding the exponents, and that is going to give me x raised to the eighth power: $(x^3)(x^5) = x^8$. He states the general rule—that the unknown raised to the mth power times the unknown raised to the nth power is equal to that unknown raised to the m plus n^{th} power ($x^m + x^n = x^{m+n}$). And the same thing is happening if you take x to the m divided by x to the n: That is going to be x to the m minus n: $(x^m)/(x^n) = x^{m-n}$.

This is also the mathematician who would be credited by al-Samawal, a 12th-century Islamic mathematician, with being the first Islamic mathematician to discover Pascal's triangle. Unfortunately, we do not have any independent evidence that this really was the first Islamic mathematician to discover Pascal's triangle. If he was, he lived before Jia Xian, and so that suggests that perhaps Pascal's triangle came up first in the Islamic world. We are uncertain about this. Whatever happened is that, amazingly, in the 11th century, Pascal's triangle appears both in China and also in the Islamic world, and it would be wonderful to know if there was some kind of communication between them. Did one of them discover Pascal's triangle and then share the information with the other, or was it just that both of them were working with this problem of arbitrary polynomials—and inevitably, as you begin to work with polynomials of arbitrary degree, you begin to come across Pascal's triangle, this idea of these powers of the binomial $1 + x$. What we find in the Islamic world, just as there was in China, Pascal's triangle comes up as a method for finding roots of arbitrary polynomials.

One of the most interesting mathematicians from the Islamic world was Abu Ali al-Haytham, born in Basra sometime around A.D. 965. Al-Haytham was an engineer, and he made his reputation in Basra, in southern Iraq, early in his career, and then sometime around the year 1000, he was enticed to come to Cairo. Cairo had been founded as the capital of the Fatimid caliphate in 969, and one of the first things that was constructed in this brand new city of Cairo was a mosque, the al-Azhar mosque. One of the first things that was done at the al-Azhar mosque was to create a school of learning. This would continue to be a continuously operating center of learning ever since its founding in 975.

In 996, al-Hakim would become the caliph. He was only 11 years old at the time, and he would become known as someone who was quite cruel and quite eccentric, but he also was someone who was very interested in developing the sciences. He went out and found good scientists to bring into Cairo to help build the scientific community in this city, and al-Haytham was one of the people that he brought in. Al-Haytham was initially brought in to be an engineer, and actually the job that he was given was to travel south along the Nile River and find a place to construct a dam so that they could control the flow of the Nile.

Al-Haytham traveled up river and discovered that the job was simply well beyond the engineering capabilities of that time. He never built the dam. When he came back to Cairo, he became rather afraid of the caliph because of his tendency both to be capricious and also cruel, and so al-Haytham feigned madness. He disappeared into his own house, but while he was doing that he was developing his mathematics and his science. One of the areas that he really worked on very extensively is optics—understanding how mirrors operate, and especially mirrors of different shapes: spherical mirrors, scientific mirrors. He would actually publish 92 scientific works—many of which dealt with optics, and many of which dealt with mathematics and other scientific topics—and 55 of al-Haytham's books still survive today.

Among the mathematics that he did was to work with the Chinese remainder theorem. Again, it is not clear if he was drawing on the Chinese tradition or if he came up with it independently. He came up with something that would later be known as *Wilson's theorem*, and that is the observation that if you have a prime number like 7, and you take the product of all of the integers that are less than that prime—so you take the product of the integers from 1 up to 6—and then add 1 more to it, and that number will always be divisible by that prime. The product of the integers from 1 to 6, plus 1, is 721, and that is divisible by 7. Eleven is a prime; the product of the integers 1 through 10, plus 1, gives you 3,628,801, and that number is exactly divisible by 11. Al-Haytham didn't just observe this—he actually was able to prove this particular result. He was using polynomials in their fully general sense, and he was able to take Archimedes's ideas for finding areas, finding volumes, and using the method of exhaustion, and he really pushed these much further than anybody had been able to do before.

One of the volume problems that Archimedes was able to solve is the problem of finding the volume of a paraboloid. So we consider a parabola, and we take the line of symmetry, and we then rotate this parabola around the line of symmetry, and that gives you a kind of bullet shape, and that bullet shape is what is called a paraboloid. Archimedes had figured out how to find the volume of a paraboloid.

The problem that al-Haytham tackled was to take this same parabola and rotate it a different way. What he did was he just took half of the parabola, and so we take the top half of the parabola, and then we take a vertical line that chops this arc of the parabola at some point, and we rotate that piece of the arc around to the vertical line, and what we get is a solid that looks very, very different from the paraboloid of Archimedes because it is a solid that comes to a point. You can actually see this solid of al-Haytham that is in the domes of many of the mosques, and this would become a common motif for Islamic mosque art.

What al-Haytham does is to work out the volume of this particular solid of revolution—obtained by rotating a parabola—but in a very different way from the way in which Archimedes did it. He does this by slicing his particular solid of revolution in much the same way that Archimedes does, but the problem is much harder. What Archimedes was able to show is that in order to find the volume of the paraboloid, you really only need to be able to find the area of a triangle, and you can reduce one problem to the other problem. For al-Haytham trying to find the volume of his solid of revolution, what he needed to be able to do was to find a formula for adding up the fourth powers of all of the integers from 1 up to n: $1^4 + 2^4 + 3^4$—all the way up to n^4.

The sum of the integers from 1 to n, that particular formula, had been known since far antiquity—and, in fact, that is so old we have no idea where it first came from. It was certainly known to the earliest Greek mathematicians, and it was known to the earliest Indian mathematicians of whom we have any record. That formula was well-known. If you consider the sum of the squares, $1^2 + 2^2 + 3^2$, up to n^2, that particular formula gives you a cubic polynomial in n, and that also is known to Greek mathematicians. The sum of cubes, $1^3 + 2^3 + 3^3$, up to n^3, that involves a fourth-degree polynomial, and we don't know when that was first discovered. It might have been discovered by the Greek mathematicians, and it was certainly known to the Indian mathematicians by the middle of the 1st millennium A.D.

Al-Haytham needs to go beyond this. He needs to add up fourth powers. One of the things that he does is find the fifth-degree polynomial that gives you the sum of the fourth powers—and beyond that, he comes up with a general procedure that can be used to find the sum of fifth powers, or sixth powers, or any powers beyond that.

I want to talk just briefly about a few of the other important mathematicians who would appear in the Islamic world after al-Haytham. The first of these is Ibn Yahya al-Samawal, who lived sometime in the middle of the 12th century in Baghdad. This was well after Baghdad's heyday; Baghdad was no longer the place to go for mathematical and scientific work, but nevertheless al-Samawal did become a very important mathematician. He was actually trained as a medical doctor. He was Jewish, and he is one of the first Jewish mathematicians that we know of from history. Eventually, in middle age, he did convert to Islam.

He worked with polynomials, and he really understood the full power of polynomials. He is someone who is able to explain how to take a polynomial and divide it by another polynomial. It is something that today we call *synthetic division*, and just as if you take one number and divide it by another, you begin to get a decimal expansion that may keep on forever. There may be no end to the decimal expansion—so, if you are going to take a number like 3 and divide it by 17, the decimal expansion repeats, but it never ends. Al-Samawal realized that if you take one polynomial and divide it by another, what you are going to get are negative powers of the unknown appearing—so you will get a term, an x^{-1}, and a term, an x^{-2}, and he realized that these negative powers could well continue forever.

He is also someone who did a lot of work with Pascal's triangle. It turns out that there is a very clever formula for finding the sum of second powers, third powers, fourth powers. Whatever power you want to fix on, you can find a formula for the sum of those powers using Pascal's triangle, using the numbers in Pascal's triangle—and al-Samawal, we believe, was the first person to discover this.

One more mathematician I want to mention just very briefly is Omar al-Khayyami (1048 to 1131), who lived in Isfahan—the very famous Persian poet. What many people don't realize is that in addition to being a poet, he also was a very accomplished mathematician. He was particularly interested in the problem of finding the exact value of the roots of a cubic polynomial, and he made progress on this but was not able to finally solve this problem. It would actually take about 400 years before this problem would be solved, and it would be solved in western Europe by Italian algebraists. This would be the first mathematical accomplishment of the European world, the western European world, and that will be the topic of my next lecture.

Lecture Nine

Italian Algebraists Solve the Cubic

Scope:

Islamic mathematics gradually spread into Europe, beginning with Leonardo of Pisa, also known as Fibonacci, in the 13th century. Fibonacci explained the Hindu-Arabic decimal system to an Italian audience in his book *Liber abaci* and worked with the sequence of integers that bears his name. Italian mathematicians began to make original contributions in the 16th century when they discovered methods for solving the general cubic and quartic equations. Such knowledge was closely guarded and used to defeat rivals in contests of mathematical prowess; thus, our story in this lecture is one of competition and intrigue. The colorful characters include del Ferro; Fontana, known as Tartaglia ("the Stammerer"); Cardano; and Bombelli. The solution of the general cubic equation would lead mathematicians to begin to work with imaginary numbers.

Outline

I. This lecture introduces us to the Italian algebraists, who learned mathematics from the Islamic scholars.

- **A.** One of the most significant insights of Islamic mathematics was the importance of the Hindu-Arabic decimal system, in which the power of 10 is determined by the place value of the digit.
- **B.** This Hindu-Arabic system of denoting numbers was used throughout the Islamic world, and by the 12th century, Italian merchants working in North Africa began to adopt it. They also learned Islamic methods in algebra, geometry, and other fields of mathematics.

II. One of the most important Italian mathematicians of this time was Leonardo of Pisa (c. 1170–1240), who referred to himself as a "son of Bonacci," or "Fibonacci," the name he is known by today.

- **A.** Fibonacci spent much of his early life in Algeria and devoted himself to learning mathematics, including algebra and geometry. Around 1200, he returned to Italy, where he published some of the most important early works in mathematics that would appear in western Europe.

B. One of Fibonacci's first books was the *Liber abaci* (*Book of Calculations*), in which he explains the base-10 system for an Italian audience.

1. The book also contains what we now call the *Fibonacci sequence*, the famous sequence of integers that begins 1, 1, 2, 3, 5, 8, 13. Each of the numbers in this sequence is the sum of the two previous numbers.

2. Although it is called the Fibonacci sequence, this sequence was known before Fibonacci by Islamic, Indian, and Greek mathematicians.

3. The Fibonacci sequence can be used to describe the number of seeds in the spirals on a sunflower head. If we follow the spirals around in two different directions, the two numbers will be consecutive Fibonacci numbers.

C. In addition to writing his book on calculations, Fibonacci also wrote about geometry, algebra, and Diophantine equations. Recall that these are equations from number theory whose solutions are restricted to integers, such as the Pythagorean triples (in which the square of an integer plus the square of another integer is equal to the square of an integer).

III. Italy, especially northern Italy and the city of Bologna, would become an important center for the study of algebra. Bologna was the site of the first great western European university, which was founded around 1088.

A. One of the mathematicians who taught at Bologna in the 15th and early 16th centuries was Scipione del Ferro (1465–1526), who was interested in the problem of finding the root of a cubic polynomial.

1. One approach to this problem is to find an approximation to the value that satisfies a given polynomial equation (i.e., an approximation to the root). The ideal solution, however, would be to find an exact value for the root.

2. If we're working with a quadratic polynomial (a polynomial of degree two), it's possible to find the exact value using square roots and the quadratic formula. Del

Ferro sought an approach for finding the exact value for the root of a cubic polynomial (a polynomial of degree three).

3. While a cubic polynomial might have more than one real root, we only need to find one root. If we find one root, then we can divide the polynomial by the *linear factor*—the factor of degree one (x minus the root)—to reduce our polynomial to a quadratic for which the quadratic formula yields the roots.
4. Del Ferro found a method for determining the exact value of one of the roots of an arbitrary cubic polynomial, and he shared his discovery with Annibale della Nave and Antonio Fiore. Del Ferro's work marked the first time in the modern western European tradition that mathematics went beyond the accomplishments of the ancient Greeks.

B. Del Ferro's discovery was also important to mathematicians who sought patrons. Those who knew how to find the roots of a cubic polynomial had a great advantage in mathematical competitions sponsored by opposing patrons.

1. One such competition took place between Fiore and a mathematician named Niccolò Fontana, better known as Tartaglia, or "the Stammerer."
2. When Tartaglia heard that del Ferro had discovered a method for finding the roots of a cubic polynomial, he began to explore the same question and figured out the method. He then challenged Fiore to a competition and won.

IV. Another mathematician who became interested in the roots of cubic polynomials was Gerolamo Cardano (1501–1576), one of the greatest Italian algebraists of the 16th century.

A. Cardano was the son of a prominent law professor in Milan. He had a reputation as a gambler and did much of his early work on probability.

B. Cardano invited Tartaglia to come to Milan and share the secret of how to find the root of a cubic polynomial. Tartaglia agreed but asked Cardano to keep the method a secret because he wanted to use it to find patronage for himself. Cardano later learned from Annibale della Nave that the "secret" was known by others.

C. In 1545, Cardano published his great work in algebra, the *Ars magna* (*Great Art*), which included the method for finding the root of a cubic polynomial.

D. At about the same time, Lodovico Ferrari (1522–1565) was sent to Cardano as a servant. Cardano recognized Ferrari's mathematical talent and made him his secretary. Ferrari quickly went beyond what Cardano had accomplished in mathematics, extending the formula for the root of a cubic polynomial to find the exact value of the root of an arbitrary polynomial of degree four.

E. Shortly thereafter, Ferrari bested Tartaglia in a competition to win an academic position in Brescia. By the end of the first day, Tartaglia was clearly falling behind and fled the competition in disgrace. Ferrari later became the tax assessor for the city of Milan.

V. Let's look for a moment at the method for finding the root of an arbitrary cubic polynomial.

A. We begin with a polynomial equation of degree three [$x^3 + \alpha x^2 + \beta x + \gamma = 0$]. If we replace x by $x - \alpha/3$, we can write this cubic equation in the form $x^3 + cx = d$. As an example, let's use the cubic equation $x^3 + 6x = 4$.

B. We use the constant, 4, and the coefficient of x, 6. We divide 6 by 3, then cube the result: $2^3 = 2 \times 2 \times 2 = 8$. Our two key numbers for this equation, then, are 4 and 8.

C. The key to finding a root of this cubic equation is to find two numbers whose difference is 4 and whose product is 8. This problem can be restated in terms of solving a quadratic polynomial, and we know how to find the exact value of the solution of a quadratic polynomial.

D. In this case, the two numbers whose difference is 4 and whose product is 8 are $2 + 2\sqrt{3}$ and $-2 + 2\sqrt{3}$. The difference of those two values is 4, and the product is 8.

E. The difference of the cube roots of $2+2\sqrt{3}$ and $-2+2\sqrt{3}$ [$\sqrt[3]{2+2\sqrt{3}} - \sqrt[3]{-2+2\sqrt{3}}$] gives the root of the original equation.

VI. Another famous Italian mathematician was Rafael Bombelli (1526–1572), an engineer born in Bologna.

A. Early in his career, Bombelli was commissioned to drain the marshes in the Val di Chiana, a high mountain valley north of Rome. The project stretched over nine years, during the course of which Bombelli read Cardano's book on algebra and began to write his own treatise on the same subject.

B. When the project at the Val di Chiana was finally complete, Bombelli was brought to Rome to take on several engineering projects for the pope, including the restoration of the Santa Maria Bridge, an ancient bridge spanning the Tiber, and the draining of the Pontine Marshes. Neither of these projects was completed successfully.

C. While he was in Rome, Bombelli discovered the *Arithmetica* of Diophantus. He began to translate this work and to incorporate it into his own book on algebra. Eventually, Bombelli produced an important work in algebra that would influence algebraists throughout western Europe.

D. Bombelli realized the importance of working with the square roots of negative numbers. Let's find the root of another cubic equation as an example: $x^3 - 15x = 4$.

1. The two key numbers that we will work with in this case are 4 and $-\left(\frac{15}{3}\right)^3$ (which equals −125). We want to find two numbers whose difference is 4 and whose product is −125.
2. We can't find two real numbers whose difference is 4 and whose product is −125, but in fact there is a root to this cubic polynomial: $4^3 - 15(4) = 4$.
3. The method of del Ferro and Cardano doesn't work in this case, but the solution can be found by working with the square roots of negative numbers. Bombelli was the first to try this approach.

4. Bombelli showed that if we allow square roots of negative numbers, then $2+\sqrt{-121}$ and $-2+\sqrt{-121}$ are two numbers whose difference is 4 and whose product is –125.
5. The cube roots of $2+\sqrt{-121}$ and $-2+\sqrt{-121}$ are $2+\sqrt{-1}$ and $-2+\sqrt{-1}$, respectively, and their difference is 4.
6. Descartes would later call these square roots of negative numbers *imaginary numbers*. In the 18th century, imaginary numbers became critical to advances in mathematics. The numbers that are built out of real and imaginary numbers, such as $2+\sqrt{-1}$, are called *complex numbers*.

VII. By the end of the 16th century, algebra was fairly well understood. In the next five lectures, we'll move into the 17th century, which is the pivot for this entire series. In Lecture Ten, we'll return to the subject of astronomy, which motivated much of the mathematics developed during the 17th century, particularly the invention of the logarithm by John Napier.

Suggested Readings:

Gindikin, *Tales of Mathematicians and Physicists*, 1–26.

Katz, *A History of Mathematics*, chaps. 8, 9.

Nordgaard, "Sidelights on the Cardano-Tartaglia Controversy."

Van der Waerden, *A History of Algebra*, chap. 2.

Varadarajan, *Algebra in Ancient and Modern Times*, 47–92.

Questions to Consider:

1. Hellenistic mathematicians worked with cubic equations, but always expressed in terms of volumes. Was the Islamic innovation of formal equations needed before the general solution of these cubic equations could be found?
2. Who, if anyone, was in the right in the Cardano-Tartaglia dispute?

Lecture Nine—Transcript
Italian Algebraists Solve the Cubic

In the last lecture we looked at the Islamic mathematicians. In this lecture we're going to look at the Italian algebraists. These are people who learned from what the Islamic mathematicians had done. One of the most important insights of the Islamic mathematicians was the significance of our Hindu-Arabic decimal system, in which you have 10 digits that can represent the different powers of 10, and so the power of 10 is determined by the place value in which you put the digit. This Hindu-Arabic system of denoting numbers would be used throughout the Islamic world, and by the 12th century, Italian merchants who were working in North Africa would begin to pick this up. They would also begin to learn about the Islamic methods of doing algebra, doing geometry, and doing other mathematics.

One of the most important Italian mathematicians of this time was Leonardo of Pisa. He sometimes wrote his name as Leonardo Pisano, and he also sometimes wrote his name referring to the family that he was from, the Bonacci family. He referred to himself as a "son of Bonacci," or "Fibonacci" in Italian. Today he is best known by this particular name, simply "Fibonacci." He spent much of his early life in Algeria. His father was actually an Italian diplomat in North Africa. We know that Fibonacci spent a lot of time learning mathematics—not just common arithmetic, but also more advanced mathematics, algebra and geometry. Around the year 1200, he returned to Italy, to Pisa. He would live the rest of his life there, and he would work on mathematics. He published some of the most important early works in mathematics that would appear in western Europe.

So we find one of the first books, the *Liber abaci*—literally the *Book of Calculations*—in which he explains for the Italian audience this strange method of representing numbers using the base-10 system, using the Hindu-Arabic numeral system. This is also the book where we find what is known as the *Fibonacci sequence*, the very famous sequence of integers that begins 1, 1, 2, 3, 5, 8, 13. Each of the numbers that we get in this sequence is the sum of the two previous numbers: So 5 and 8 is 13, and 8 and 13 is 21, and 13 and 21 is 34, so we get this sequence. Fibonacci was actually looking at this sequence in order to count the number of rabbits you would have after a certain number of generations.

Although we call it the Fibonacci sequence, it actually is much older. It's a sequence that was known to Islamic mathematicians, to Indian mathematicians. You can even trace it back to the Greek mathematicians of the classical age. They certainly were aware of this particular sequence, and it's a fascinating sequence. There are all kinds of things you can do with it. One of the classic examples is looking at the seeds in a sunflower head. If you follow one of the spirals, the number of seeds in a given spiral is always going to be a Fibonacci number. You can actually follow the spirals around in two different directions, and the two numbers are going to be consecutive Fibonacci numbers. In addition to writing his book on calculations, Fibonacci also wrote on geometry, he wrote on algebra, and he wrote on Diophantine equations. These are equations from number theory in which you're only looking for integer solutions—things like the Pythagorean triples—where the square of an integer plus the square of another integer is equal to the square of an integer.

Italy, especially northern Italy, would begin to become an important center for the study of algebra. Most of the study of algebra actually was centered on the city of Bologna. Bologna was the site of the first of the great western European universities. It was founded before the University of Paris, before Oxford or Cambridge. It was around 1088 that the university was first founded in Bologna, and one of the mathematicians who taught there in the 15th century and early 16th century was Scipione del Ferro. Del Ferro was interested in the problem of finding a root of a cubic polynomial, so this is a polynomial of degree three.

I need to say a little bit about finding the roots of a polynomial. You've got a polynomial equation; you're trying to find out where that is equal to zero. In the last lecture I talked about some of the Islamic methods—and before that, Chinese and even Indian methods—for finding roots of polynomials. One approach is to try to find an approximation to the value that satisfies a given polynomial equation—in other words, trying to find an approximation to the root.

Ideally, what we'd like to do is to find an exact value for that root, the exact number that satisfies it. If we're working with a quadratic polynomial (a polynomial of degree two), then it is possible to find the exact value using square roots and what's commonly referred to as the *quadratic formula*. That gives you an exact value for the root.

This is the question that del Ferro was pursuing. Is there a way of finding an exact value for a root of a cubic polynomial? Now today when we think of cubic polynomials, it's degree three, and usually we want to write down all three roots. But in fact, you only really need to be able to find one of those roots, because if you can find one root of a given polynomial, then you can divide that polynomial by the *linear factor*—by the factor of degree one (x minus that root)—and you're going to get a polynomial of one lower degree. So, if I've got a polynomial of degree three, and I can find one root for that polynomial, I can then translate that polynomial into a polynomial of one lower degree, a quadratic polynomial—and now there's standard methods for finding the roots of the quadratic.

So the real problem is just to find one root of the cubic polynomial. Once you do that, you've simplified the problem down to finding the root of a quadratic polynomial (or a polynomial of degree two). Del Ferro succeeded in doing this, and he shared his discovery with two other people. One was a colleague by the name of Annibale della Nave, and della Nave is going to appear again later in this story. Another person that he shared the secret of how to find a root of a cubic equation with was Antonio Fiore. Finding this root of the cubic polynomial, of an arbitrary cubic polynomial, would turn out to be extremely important because this is really the first time in the modern western European tradition that mathematicians actually were able to go beyond what the ancient Greeks had accomplished. So this was really showing, this now happening in the early 1500s, that European mathematicians of this time were actually able to go beyond what the ancients had managed to accomplish. Finding the root of a cubic polynomial was a very important watershed.

It also was important for the individuals who were able to do that because at this time, if you wanted to be a scientist or a mathematician, you needed to have a patron. You needed to have a sponsor. You really couldn't survive on your own. You could teach some math classes, but that didn't bring in very much money. One of the ways to get a sponsor was to engage in competitions. So one patron would take his mathematician, and another patron would take his mathematician, and the two mathematicians would have a competition. Whichever patron had a mathematician that was able to win—well, certainly that brought great glory to the given patron. Mathematicians who knew how to find the roots of a cubic polynomial now suddenly had a great advantage in these competitions.

One of the important early competitions took place between Fiore and another mathematician by the name of Niccolò Fontana. Fontana is better known as "Tartaglia" ("the Stammerer"), and he got his name because as a young man in Brescia, where he grew up, one day while he was quite young, the French corsairs came through the city. They were raiding the city, and Fontana was slashed with a saber, and he suffered quite a deep gash across his face. Apparently it never healed properly, and so he had a speech impediment for the rest of his life. From that, he developed the nickname "the Stammerer" or "Tartaglia." Fontana (or Tartaglia) would turn out to be quite unfortunate throughout his entire lifetime, but he was a brilliant mathematician. When he heard that del Ferro had discovered how to find the roots of a cubic polynomial—or to find one root of a cubic polynomial—he began to explore this question for himself, and he discovered how to do it.

He then challenged Fiore to a competition. Fiore had learned how to find the roots of a cubic polynomial, but Fiore was not a great mathematician. This was basically his entire bag of tricks. On the other hand, Tartaglia was quite accomplished. He had figured out how to find the roots of a cubic polynomial, but he also knew how to solve many other mathematical problems. So when they held this competition, Tartaglia posed many different problems—not just finding roots of cubic polynomials—while Fiore only posed the question of finding roots of cubic polynomials. Tartaglia went away from the competition with Fiore with a great deal of glory because he was able to answer all of Fiore's questions, and Fiore was able to answer none of Tartaglia's.

Another mathematician who then became interested in this question of the roots of the cubic polynomial was Gerolamo Cardano—born in 1501, would die in 1576—certainly one of the greatest of the Italian algebraists of the 16th century. He was from Milan. He was the son of a very prominent law professor, scholar of law, in Milan. Somewhat spoiled, he had quite a reputation as a gambler, and that tied into his mathematics. He actually did a lot of early work on probability. He also became known as a quite accomplished mathematician, but the one thing he was not able to do was to figure out how to find the root of an arbitrary cubic polynomial. So he wrote to Tartaglia and asked Tartaglia to share the secret of how to find the root of a cubic polynomial. He invited Tartaglia to come to Milan. Tartaglia did so, thinking that Cardano might be able to find a patron for him. Despite his success in this competition against Fiore, Tartaglia had not been able to find himself a patron, and so he went to Milan hoping that Cardano could help him with this.

Well, he went there, and he shared the secret of how to find the root of a cubic polynomial with Cardano. Cardano, however, did not find a patron for him at the time. Tartaglia asked Cardano to keep this a secret because Tartaglia wanted to use this in order to arrange for patronage for himself. Cardano said: Okay, but please publish it. I would like to be able to talk about this method for finding the root of a cubic polynomial. Tartaglia went home, hoping to get this patronage, which never came. Cardano was waiting in Milan for Tartaglia to actually publish the result so that Cardano could then publish more about it himself—so both were waiting for the other.

Eventually, Cardano went to Bologna, and there he met della Nave. Della Nave was that colleague of del Ferro that I talked about earlier in this lecture. Cardano learned from della Nave that in fact this "secret" of how to find the root of a cubic polynomial was already known, had been known long before Tartaglia. Cardano, up to that point, had only known about Tartaglia's discovery of how to find the root of a cubic polynomial. Once Cardano realized that there were others who knew about this, he felt that he had been released from his promise. So he went home, and he now published his great work in algebra, the *Ars magna*—literally, the *Great Art* in 1545. In that book, despite his promise to Tartaglia, he explained the method for finding the root of a cubic polynomial, which won him the eternal enmity of Tartaglia.

At the same time, there was another young potential mathematician growing up in Bologna, Lodovico Ferrari. Ferrari was born in 1522 and would die in 1565. Ferrari's father died when he was quite young, and he went to live with his uncle. One of his uncle's sons, Ferrari's nephew, had traveled to Milan and gotten a job as a servant to Cardano. Apparently Ferrari's nephew didn't like being a servant for very long, and after a few months he simply left without giving Cardano any warning. Cardano was quite upset, and so he wrote to the young man's father and said, "You must send him back. I need him as a servant." The father, Ferrari's uncle, was reluctant to force his son to go back, and so he took his nephew, the young Lodovico Ferrari, and said that he had to go back and serve as a servant to Cardano in his place.

This actually turned out to be a great stroke of luck for Lodovico Ferrari because he went to Cardano, and Cardano immediately recognized that this man had a great deal of mathematical talent. Rather than being simply a servant to Cardano, Ferrari quickly became a secretary, and Cardano undertook to teach him the mathematics that he knew. Ferrari was very accomplished in this and very quickly went beyond what even had been done by Cardano. He was able to take a look at the way of finding the root of a cubic polynomial—an exact formula for the root of a cubic polynomial—and expand this idea so that he was able to find the exact value of the root of an arbitrary polynomial of degree four. So he went from polynomials of degree three to degree four. Shortly after that, an academic position with a patronage opened up in Brescia, and Ferrari decided to apply for this position. Tartaglia was still stewing. He was in Venice at this time, and he decided also to apply for this position in Brescia.

In order to decide who would get the position, a competition was arranged between Ferrari and Tartaglia. Ferrari was such an accomplished mathematician that in the first day of this—and one can imagine what it must have been like. There would have been a stage set up in the plaza with the two mathematicians facing each other, and one would propose a problem, and the other had to work it out in front of the crowd. Then if he did it successfully, he would then propose a problem to the other one, who then had to work it out in front of the crowd. It must have been a great scene. But at the end of the first day, it was quite clear that Tartaglia was falling behind. There were a lot of Ferrari's posed problems that Tartaglia was unable to solve, but Ferrari was able to solve all of the problems that Tartaglia posed.

So Tartaglia left before the competition was over. At the end of the first day, he fled in disgrace. He had clearly lost the competition. Ferrari was exultant that he had succeeded. He was entitled to this position with patronage in Brescia. He had bested Tartaglia. His fame was such that he was able to take any job, essentially, that he wanted in north Italy, and so he went back to Milan and got a job as the tax assessor for the city of Milan, a position that I'm sure was extremely lucrative. He would continue to do mathematics, but he quickly became a very wealthy man.

Now before I go on to the next mathematician I want to talk about, who is Bombelli, it's important to say a little bit about the method for finding the root of an arbitrary cubic polynomial. I've got a polynomial equation of degree three, so I've got some polynomial that involves x^3, and I want to find an exact value for one of the roots. Now it's always possible to translate the variable (to move the variable) so that you can write this cubic equation in the form x^3 plus some number times x is equal to a constant number. As an example, we might take the cubic polynomial $x^3 + 6x = 4$. We now take two numbers—we take that constant number 4, and we take the coefficient of the x. I had $x^3 + 6x = 4$. I take that 6, and I divide it by 3. That gives me 2, and then I cube that number. I multiply three copies of that number together: $2 \times 2 \times 2 = 8$.

So my two key numbers for this particular cubic equation are 4 and 8. The key to finding a root of this cubic equation is to find two numbers whose difference is the 4 and whose product is the 8. Now this is an old, old problem. It goes right back to the Babylonians. You know the difference between two numbers. You know the product of two numbers. You want to find out what those two numbers are. That is a problem that can be restated in terms of solving a quadratic polynomial, and we know how to find the exact value of the solution of a quadratic polynomial. In this particular case, the two numbers whose difference is 4 and whose product is 8 are $2 + 2(\sqrt{3})$ and $-2 + 2(\sqrt{3})$. So if I take the difference of those, I get 4; if I take the product of those, I get 8. I take those two numbers that I have found—$2 + 2(\sqrt{3})$ and $-2 + 2(\sqrt{3})$—and I take the difference of their cube roots, and that will give me a root of the original equation.

Now the next mathematician I want to look at is Rafael Bombelli. Bombelli was born in 1526 in Bologna. He actually worked as an engineer. He was not trained as a mathematician. He did not intend to be a mathematician. Early on in his career, he was commissioned to go up to the Val di Chiana—this is a high mountain valley that is north of Rome—and to drain the marshes there. Today the Val di Chiana is a very fertile land, a very important agricultural area, but at the time that Bombelli went up there, it was all marshland. So he was sent as an engineer to drain those marshes, and he succeeded in doing it, but it wound up taking him nine years to get these marshes drained. The reason for it is in the middle of that nine-year period, there was a five-year hiatus in which there was a dispute over who actually owned the land that was in that particular valley.

So, Bombelli was up there in the Val di Chiana cooling his heels, waiting for word to come through to begin to proceed with the work once the dispute over who owned the land was over. During that five-year period, he was reading algebra. He was reading the work of Cardano, and he was very impressed with what Cardano had done, but he decided that he could do an even better job, and so he began to write his own book on algebra. Well, once the marshes at the Val di Chiana were drained, Bombelli had made his reputation as an engineer, and he was brought to Rome by the pope for a couple of engineering projects that the pope had in mind.

The first of these was to restore the Santa Maria Bridge. This is one of the ancient bridges of Rome. It spans the Tiber at a very important crossing point, and periodically it had been destroyed over the years by various floods. After one of these floods, the pope brought Bombelli in to try to restore the bridge. Ultimately, Bombelli was not successful, and I do have a picture of what the Santa Maria Bridge looks like today. As you can see, it's not very appropriate for trying to get across the Tiber. People finally abandoned any attempt to restore the Santa Maria Bridge in 1598, and today it's simply known as the *Ponte Rotto*, the "Broken Bridge." Another one of the jobs that the pope had for Bombelli was to try to drain the Pontine Marshes. This, also, Bombelli was not successful in, and actually the Pontine Marshes, which are a great coastal marshland south of Rome, would not be drained until the 1920s. It would be under Mussolini that eventually these marshes would be drained.

While he was in Rome, Bombelli discovered Diophantus's *Arithmetica*. This is the great work on number theory and Diophantine equations that Diophantus had written in the early centuries A.D. He began to translate this work and to see how to incorporate it into his own book on algebra. So eventually, Bombelli would produce a very, very important work in algebra that would go on to influence not just the Italian algebraists of the succeeding years, but also algebraists throughout all of western Europe. One of the most important insights that Bombelli had was the importance of being able to work with the square root of negative numbers.

Let me take another cubic equation that I want to find the root for. I'm going to take $x^3 - 15x = 4$. The two numbers that we're working with in this case are 4 and $-\left(\frac{15}{3}\right)^3$. So I want the difference between two numbers to be 4, and I want the product of two numbers to be −125. Now that is a problem because if I've got two numbers that are only 4 apart from each other, and the product of their magnitudes is 125, either these numbers both have to be positive or both have to be negative. I can't get a −125. You can't find two real numbers whose difference is 4 and whose product is −125. But in fact, there is a root to this cubic polynomial. It's not too hard to see that $4^3 - 15(4) = 4$; 4 is a solution.

So, Bombelli wrestled with this question. The method of Cardano, and Tartaglia, and del Ferro doesn't work in this case, but there is a solution. What is going on? It turns out that the key to this solution is to be able to work with the square roots of negative numbers, and Bombelli is the first person to start doing this. What he does is he shows that you get two solutions, one of which is the two numbers whose difference is 4 and whose product is −125. The numbers are $2 + \sqrt{(-121)}$ and $-2 + \sqrt{(-121)}$. If you take the cube roots of those numbers, you get $2 + \sqrt{(-1)}$ and $-2 + \sqrt{(-1)}$. Well, $(2 + \sqrt{(-1)}) - (-2 + \sqrt{(-1)})$, take the difference—you get 4; you get the solution. You get the solution, but on the way you need to work with these square roots of negative numbers.

So, Bombelli is the first person to say, yes, we do need to work with the square roots of negative numbers. The square roots of negative numbers would later be called by Descartes *imaginary numbers*. That is a very unfortunate name. Descartes called them imaginary, and a lot of people seem to think that they don't exist because we call them imaginary. They are very important numbers. They're very real numbers in some sense, although we refer to the ordinary numbers as "real."

As we're going to see several lectures hence, in the 18th century these imaginary numbers become absolutely critical to proceeding with mathematics, to making advances in mathematics. The numbers that can be built out of real and imaginary numbers, the *complex numbers*, are really two-dimensional numbers, and I'll talk about these much more later. We've reached the end of the 16th century. Algebra at this point is pretty well understood. In the next lecture we're going to be moving into the 17th century, the century that I said is the pivot for this entire series of lectures. We're going to spend five lectures on the 17th century. These developments in algebra undertaken by the Italian algebraists would lay the foundation for this.

In the next lecture we're going to go back to questions in astronomy. Astronomy led us into trigonometry, and in the next lecture we're going to see some of the important ways in which astronomy would provide the motivation for the mathematics that would be developed during the 17th century. In particular, we're going to focus on John Napier, a Scotsman, and his invention of the logarithm.

Lecture Ten
Napier and the Natural Logarithm

Scope:

In this lecture, we look in detail at logarithms, a tool invented to assist astronomers in making accurate calculations from their observational data of the heavens. Two astronomers in particular, Tycho Brahe and Johannes Kepler, sought to improve Copernicus's earlier model of the universe, in which he posited that the planets move around the Sun in circles. Kepler eventually discovered that the paths of the planets are elliptical rather than circular. John Napier was a Scottish nobleman who became interested in finding a computational tool that would exploit the power of exponents to facilitate astronomical calculations. He created tables of logarithms that were accurate to 7 digits, a remarkable result equivalent to calculating the first 23 million powers of 1.0000001. Napier's logarithms would later play an important role in calculus.

Outline

I. This lecture explores the most important functions to be discovered since the trigonometric functions—the logarithm and the exponential function. Before we look at these topics, however, we need to see what was happening in astronomy in the middle of the 16th century because developments at that time would influence 17th-century mathematics.

II. The year 1543 saw the publication by Nicolaus Copernicus (1473–1543) of his groundbreaking work, *On the Revolutions of the Heavenly Spheres*.

- **A.** Copernicus was a Catholic monk in Poland and was interested in the motion of the planets. In particular, he was disturbed by the problem of retrograde motion that we discussed in Lecture Five—that is, the apparent backward motion of the planets at certain times.
  - **1.** Archimedes had raised this problem, and Aristarchus had suggested the idea of the Earth traveling around the Sun as an explanation of retrograde motion. That explanation was dismissed because we on Earth have no sense of the planet's motion.

2. Slightly later, Apollonius of Perga offered the theory of epicycles—that planets move in circles whose centers then travel around the Earth—to explain retrograde motion.
3. This explanation became the foundation for what was known as *Ptolemaic astronomy*, named after the Greek astronomer Ptolemy, who showed how the movement of the planets could be modeled using epicycles and equants.
4. Over the succeeding centuries, the system became more and more complicated because it never quite matched what people observed in the heavens. Epicycles were added to epicycles.

B. Copernicus sought to simplify the model by going back to the idea of Aristarchus's. He postulated that, in fact, the Earth travels around the Sun. The problem with Copernicus's work was that he also assumed that the planets travel around the Sun in a circle, but observations were sufficiently accurate by this time to show that this model does not accurately predict the locations of the planets.

III. To improve the model, more accurate observations of the positions of the planets were needed; the man who undertook to make these observations was a wealthy Danish astronomer named Tycho Brahe (1546–1601).

A. Brahe, who had studied astronomy in Copenhagen and Germany, set up an astronomical observatory on the island of Hven, between what are now Sweden and Denmark.

1. Brahe made two important observations in Hven. The first, in 1572, was the observation of a supernova, which revealed that the stars are not eternal. This discovery undermined the Aristotelian view of the heavens.
2. Brahe also observed and measured the path of a comet in 1577. Comets were once believed to be phenomena that were confined to the atmosphere of the Earth, but Brahe showed that the comet he observed was well beyond the Moon. Again, the fact of something moving through the heavens revealed that the universe is not perfect and unchanging.

B. Brahe made all his observations without a telescope. The best one can usually hope for in this case is an accuracy of about 2′ of arc (360° is a full circle, and $\frac{1}{60}$° is a minute). Differentiating between two stars in the sky separated by less than 2′ of arc is difficult, but Brahe was sometimes able to fix the positions of the planets to within 1′ or even $\frac{1}{2}$′ of arc.

C. In 1599, Brahe was hired as the court astrologer to the Holy Roman Emperor Rudolf II, whose capital was in Prague. Shortly after he took the position, he hired a young assistant, Johannes Kepler (1571–1630).

1. In 1601, Brahe died from what was later determined to be mercury poisoning.
2. It has been suggested that Kepler had a hand in Brahe's death. He benefited from it, inheriting both Brahe's position as court astrologer and his valuable data.

D. It became Kepler's life work to find a model that would fit these meticulous data. Eventually, he realized that the reason the model of Copernicus's was off is that the orbits of the planets are not circles but ellipses, and the Sun is located at one focus of the ellipse.

E. Kepler also established other laws of celestial motion. He realized that the planets sweep out equal area in equal time. This means that the planets speed up when they come close to the Sun and slow down when they move away.

1. Another one of Kepler's laws is that the square of the time required to complete an entire revolution of the Sun is proportional to the cube of the average distance to the Sun.
2. In many respects, Kepler was an astrologer and a mystic. He explained the relative distances of the planets in terms of Platonic solids, each one sitting inside another. He believed that the crystalline spheres in which the planets were embedded rubbed against each other, and he described the music of the spheres.

IV. With the data he inherited from Brahe, Kepler needed to perform calculations that were accurate to at least 5 digits.

A. Shortly after Kepler began working with Brahe's data, the telescope was invented, and Kepler was able to accumulate even more accurate data. Soon, astronomical calculations required 7 digits of accuracy, and by 1630 they required 10 digits of accuracy.

B. The Scottish nobleman John Napier (1550–1617) invented a computation device to assist in performing multiplication and division and evaluating trigonometric functions to 7 or 10 digits of accuracy.

1. Napier had wide-ranging interests. One of his most famous writings was a diatribe against the Catholic Church, in which he theorizes that the revelation of St. John depicts the pope as the Antichrist.
2. Napier traveled to Italy in 1594 and may have met Galileo. He returned from Italy with an interest in developing computational tools to facilitate difficult multiplication and division problems.
3. In 1614, Napier published his work, *Description of the Wonderful Canon of Logarithms*.

C. Napier may have invented the word *logarithm* by combining two Greek words, *logos* and *arithmos*. As mentioned in an earlier lecture, the Greeks had separated mathematics into two different spheres; one was *logos*, or logical mathematical reasoning, and the other was *arithmos*, or calculation. The beauty of the logarithm is that it applies logical mathematical reasoning to calculation.

D. The idea behind the logarithm is to exploit the power of exponents. Let's look at a simple example.

1. If we want to multiply 16×128, we know that 16 is 2^4 and 128 is 2^7. Earlier Islamic mathematicians had known that multiplying $2^4 \times 2^7$ can be accomplished by adding the exponents: $2^{11} = 2048$.
2. What if we want to multiply 19×33? There is a power of 2 that gives us 19; it falls between 2^4, which is 16, and 2^5, which is 32. The exact value is 4.247928.
3. In the same way, we can write 33 as a power of 2, and we can multiply 19×33 by adding the two exponents. The result is $2^{9.292322}$, which is approximately 627. Because 19×33 is an integer, we know that it will be exactly 627.

4. We started with 19, and we wanted to find the exponent of 2 that would give us 19. Napier called the exponent the *logarithm* of 19. The number that we exponentiated is called the *base*. Thus, 2 is the base, and 4.247928 is the logarithm of 19 for the base 2.
5. The problem then becomes: Given an integer, find the exponent of 2 that will lead to that integer; in other words, find the logarithm of that integer for the base 2.
6. This introduces the idea of the exponential function. We're not just interested in the integer powers of 2 or the simple fractional powers of 2. We're now interested in 2 raised to a power that could be any real number.

E. One way to calculate logarithms to an accuracy of 7 digits is to calculate the first 23 million powers of 1.0000001.
   1. Napier realized that performing those calculations was an impossible task, but his insight that a simple conversion factor could be used to find logarithms for any base once they were known for another base enabled him to find the 23 million values easily.
   2. He found the first 230 powers of 1.01, which is roughly equivalent to finding 230 powers of 1.0000001 that are multiples of 100,000. He then interpolated between those, finding powers that are multiples of 2000, then multiplies of 100.
   3. With just a few hundred calculations, Napier was able to set up tables that would enable mathematicians to find logarithms to the base 1.0000001 for all numbers between 1 and 10, yielding the desired 7-digit accuracy.

F. Because Napier could change bases so easily, he realized that there was no need to stick with the base 1.0000001. The question now became what base to use.
   1. Consider two lines with points moving along each line. On the first line is the point that we want to take the logarithm of. On the second line, we look at the image of that point under the logarithm.
   2. For example, for the point 3 on the first line, the position on the second line is the logarithm of 3. When the point moves to 5 on the first line, the point on the second line moves to the logarithm of 5.

3. If the point on the first line travels at a uniform speed, how fast does the point on the second line move? Napier saw that the point on the second line moves at a speed that is inversely proportional to the distance that the point on the first line has moved.
4. In other words, if the first point has moved to position x, the speed of the point on the second line is some constant divided by x. The value of the constant depends on the base that we are using for the logarithm.
5. Napier chose the logarithm that would give him a speed of exactly $1/x$. Today, we call this the *natural logarithm*.
6. Napier never worked out the value of this base, but it was later observed to be about 2.71. This number is referred to as e, which is called the *base of the natural logarithm*.
7. Another mathematician, Henry Briggs, in London, realized that a base of 10 would be much easier to work with. He would publish his own table of logarithms that was accurate to 10 digits.
8. Napier's logarithm, however, would have an important role to play in calculus. Two Belgian Jesuits discovered that the area underneath the curve of a hyperbola from 1 to any value t is given by Napier's natural logarithm of t.

V. In the next lecture, we'll return to the early 1600s to look at the work of Galileo and the beginnings of the mathematics of motion.

Suggested Readings:

Katz, *A History of Mathematics*, 416–20.

Toeplitz, *The Calculus*, 86–94.

Whiteside, "Patterns of Mathematical Thought," 214–31.

Questions to Consider:

1. In what ways do Napier's logarithms combine the two Greek conceptions of mathematics, *logos*, or logical reasoning, and *arithmos*, or calculation?
2. The natural logarithm fascinated scientists of the 17th century because it gave them another example of what today we would call a function, a well-defined rule for which each input value

produces a unique output value. What is special about this function that distinguishes it from the other functions, especially the trigonometric functions, that were known at the time?

Lecture Ten—Transcript

Napier and the Natural Logarithm

In this lecture we're going to move into the 17th century and look at the most important function to be discovered since the trigonometric functions—namely the logarithm, and also the exponential function. Exponentials were certainly dealt with before the 17th century, but as a function that really only begins in the 17th century.

But, before I start to look at logarithms and exponentials, I need to back up a bit into the middle of the 16th century and look at what was happening in astronomy, because that would directly influence what would then happen in the 17th century. The year 1543 is the year that Nicolaus Copernicus published his groundbreaking work, *On the Revolutions of the Heavenly Spheres*. Copernicus was a Roman Catholic monk in Poland, and he was trying to understand the motion of the planets. In particular, he was very disturbed by this problem of retrograde motion, this apparent backward motion of the planets at certain times.

If you'll recall, back when we looked at astronomy in ancient Greece, this was one of the problems that Archimedes raised, and a number of different Greek scientists, astronomers, looked at this problem. One of them was Aristarchus, who suggested that what was happening was that the Earth was actually going around the Sun, and that would explain this retrograde, or backward, motion of the planets. The problem with that explanation of Aristarchus was that we have absolutely no sense of a motion of the Earth. It appears that the Earth is stationary, and so that was quickly dropped in favor of a slightly later explanation that came from Apollonius of Perga, the same Apollonius who did the important work on conics. It's Apollonius who came up with the idea of epicycles—that, in fact, the Earth is stationary at the center of the universe, and what is actually happening is that the planets are traveling around the Earth on little circles whose centers then travel around the Earth.

This would become the foundation for what was known as *Ptolemaic astronomy*. Certainly there were many Greek astronomers before Ptolemy who established this system, but Ptolemy really showed how you could model the movement of the planets using these epicycles. Over the succeeding centuries, the system became more and more complicated because whatever system was in place never really matched what people were observing in the heavens. So they tried the idea of equants. They tried putting epicycles on epicycles. They got very complicated models. What Copernicus decided to do was to simplify this process, or try to simplify the process, by going back to the idea of Aristarchus and postulating that it was in fact the Earth that goes around the Sun, rather than the Sun that goes around the Earth.

The problem with Copernicus's work is that he also assumed that everything went around the Sun in a circle. Observations were sufficiently accurate by this time, the mid-1500s, that that simply did not accurately predict where the planets should be. The planets weren't where they were supposed to be if in fact they were traveling in nice circles around the Sun. So, a number of astronomers began to try to improve on this model and figure out what was really happening. Something else with Copernicus is that in his book, *On the Revolutions of the Heavenly Spheres*, actually he probably did believe that the Earth moves, but he simply puts it forward as a hypothesis, as a possible model. But succeeding astronomers began to suspect that maybe, in fact, the Earth is moving around the Sun.

What was needed were much more accurate observations of the positions of the planets on which a better model could be built. The person who undertook to find these observations was a Danish astronomer by the name of Tycho Brahe, born in 1546; he would die in 1601. He was a very wealthy nobleman. It has actually been estimated that at one point he personally owned 1% of all of the wealth in Denmark. He set up an astronomical observatory on the island of Hven. It's just off the coast of Sweden, actually between what is now Sweden and what is now Denmark, although at that time what is now southern Sweden was part of Denmark. Brahe had studied astronomy in Copenhagen. He had then gone to Germany to continue his study of astronomy. While in Germany, he engaged in a duel where his nose was sliced off. Rather unfortunately, he went through the rest of his life with a tin nose, something that makes him very distinctive. We know that he also studied medicine and alchemy.

He made two important observations in his observatory in Hven. The first (1572) was a careful observation of a supernova. A star suddenly becomes a brilliant light in the sky and then disappears. Brahe observed this, and one of the things that this showed is that, in fact, the stars in the heavens are not eternal. They can appear out of nowhere; they can disappear. So this began to undermine the Aristotelian view of the heavens. Something else that Brahe observed in 1577 was a comet. Now if you really believe that the heavens are perfect, you can't have anything moving through the heavens. It was believed up until that point that comets were actually something that happened within the atmosphere of the Earth. Brahe did very careful observations and measurements of the position of this comet, and he was able to show definitively that the comet, in fact, was well beyond the reach of the Moon. It was well outside the atmosphere of the Earth, and so it really was moving through the heavenly reign.

Brahe, you must remember, was working before the time of telescopes, and so all of the observations that he was making were being made simply by the naked eye—with very accurate instruments to line things up, but without the aid of a telescope. Now the best you can normally hope for in this case is to get an accuracy of about 2′ of arc. (So, 360° is a full circle, and (1/60)° is a minute). Actually, if there are two stars in the sky whose difference is less than 2′ of arc, it's very difficult to tell them apart, so that's normally about the best you can do with naked-eye observations. Brahe actually was able to push that a little bit further. Looking back at his records, we now realize that sometimes he was able to get the position of the planets down to within (1/2)′ of arc, very often within 1′ of arc. He had these incredibly accurate observations.

In 1599, he was hired by the holy Roman emperor of the time, Rudolf II (whose capital was in Prague), to be the court astrologer. As I mentioned in earlier lectures, there really at this time was no difference between a mathematician, an astronomer, and an astrologer. Brahe, as an accomplished astronomer, would also have been much sought after as an astrologer—trying to predict the position of the planets and interpret what that actually meant for what would go on on Earth. Shortly after he became court astrologer, he hired a young assistant, Johannes Kepler. Kepler had been born in 1571.

Shortly after Kepler became Brahe's assistant, Brahe died. There was a story circulated at the time that the reason that he died was that he was at a banquet with the emperor, and he desperately needed to relieve himself. But being seated at table with the emperor, he could not simply get up and leave the table until the emperor was ready to get up and leave the table. The story was that his bladder burst, and he died of the infection. About 50 years ago, Brahe's corpse was exhumed and examined. I'm not sure how much of his bladder was left at that point, but they did an analysis of his hair and discovered that, in fact, he had died of mercury poisoning. Now this raises all kinds of suspicions because, in fact, Kepler benefited greatly from Brahe's death, and it has been suggested that perhaps Kepler had a hand in doing in Brahe. Kepler inherited Brahe's position as court astrologer to Rudolf II, and he also inherited all of Brahe's very valuable data.

It became Kepler's life work to take this meticulous data and try to find a model that would fit it. Kepler eventually would succeed in doing this, but to do it he needed to do very extensive calculations, very involved calculations. Eventually, he would come to the realization that the reason that Copernicus was off is that the orbits of the planets are not circles—they are actually ellipses. They're ellipses not with the Sun at the center of the ellipse, but with the Sun off at the focus of the ellipse. He realized that the data that Brahe had collected could be explained with this model. Kepler came up with other rules, other laws of celestial motion. He realized that the planets sweep out equal area in equal time. Practically, what this means is that the planet speeds up as it comes in close to the Sun, and then it slows down as it moves away from the Sun, so it's not always traveling at the same speed.

Another one of Kepler's laws is that the square of the time that it takes to complete an entire revolution is proportional to the cube of the average distance to the Sun. Kepler came up with many other laws. In many respects, he really was an astrologer and a mystic. He explained the relative distances of the planets in terms of Platonic solids, each one sitting inside another. He also talked about the way in which the crystalline spheres that the planets were embedded in would rub against each other. He talked about the music of the spheres, and he actually describes the kinds of harmonies that the planets make as they travel around the Sun.

But I want to get back to his question of trying to work with building this model by doing these very involved calculations. With the kind of data that he had inherited from Brahe, Kepler needed to be able to do calculations that were accurate to 5 digits. In fact, shortly after

Kepler began doing his work on Brahe's data, the telescope was invented. Galileo would be the first person to actually begin to use the telescope to look at the stars, and this happened very early in the 1600s. Kepler then would begin to get even more accurate data. Very quickly, he needed not just to do 5-digit calculations, but 7-digit calculations. Certainly by 1630, the data that was available coming from the telescopes was sufficiently accurate that people needed to be able to do calculations that maintained 10 digits of accuracy.

Doing multiple multiplications and divisions, and evaluating trigonometric functions to 5, or 7, or 10 digits of accuracy, is extremely difficult, and so a computational device was needed in order to make this easier to do. That computational device would be invented by the Scotsman, John Napier. Napier was born in 1550. He was a Scottish nobleman. He was the Laird of Merchiston in what today is now part of Edinburgh. He had many interests. He was interested in theology. One of his famous works was published in 1593, and it was a diatribe against the Catholic Church called the *Plaine Discovery of the Whole Revelation of St. John*, in which he talks about how the revelation of St. John is really describing the pope as the Antichrist.

But despite this great skepticism of the Catholic Church, he did travel to Italy in 1594. It's very possible that he actually met Galileo at that time. He clearly came back from Italy with an interest in developing these computational tools for making it easier to do multiplications and divisions. In 1614, he finally published his work, what he called *Description of the Wonderful Canon of Logarithms*. The word *logarithm* is a word that was invented by Napier to describe what he had created, and there are various ways of understanding its etymology. You can actually see its root in the phrase "ratio number," which is a reasonable explanation for what is going on with a logarithm.

But the etymology that I really like is to take logarithm as being composed of two Greek words: *logos* and *arithmos*. If you'll recall back when I was talking about Greek mathematics, I said that they viewed mathematics as two different areas: One was *logos* or logical mathematical reasoning; and the other was *arithmos*, from which we get our word "arithmetic," and this is calculation. The beauty of the logarithm is that it applies this logical mathematical reasoning to the problem of calculation, and that is precisely what Napier was getting at, trying to simplify calculations.

The idea behind the logarithm is to exploit what happens when you're working with exponents. Let me give you a simple example. Let's say that I want to multiply 16×128. Well, if I know my powers of 2, I know that $16 = 2^4$, and I know that $128 = 2^7$. Now, one of the results that the Islamic mathematicians were certainly very well aware of—and was now known in western Europe—is that if I want to multiply $2^4 \times 2^7$, I can accomplish that by adding the exponents. So, $2^4 \times 2^7$—that is four copies of 2 times another seven copies of 2—I get 11 copies of 2: $4 + 7 = 11$. If I know my powers of 2, I can simply read off a table: $2^{11} = 2048$. Well, that is great if I'm trying to multiply two powers of 2.

What if I want to multiply 19×33? Well, it is possible to write 19 as a power of 2. You can work with fractional powers of 2: $2^{½} = \sqrt{2}$; $2^{5/3}$ equals 2^5, and then take the cube root of that. So, if I have any rational exponent that actually represents taking a power and then taking a root, that represents some number. There actually is a power of 2 that gives us 19. It's going to be between 2^4, which is 16, and 2^5, which is 32. The actual power of 2 that gives you 19, to 7-digit accuracy, is 4.247928. In the same way, we can write 33 as a power of 2, and then we can multiply 19×33 by adding these two powers of 2. That gives us $2^{9.292322}$. If I've got a table of powers of 2, I can see that that's approximately 627. Since 19×33 is an integer, I know that it's got to be exactly 627.

Now if I look at that exponential equation, $2^{4.247928} = 19$, what is happening there—if I start with the 19 and I want to know what that exponent is, what Napier did was to call the exponent the *logarithm* of 19. The number that we're exponentiating that to is called the *base*. So, 2 is the base, and 4.247928 is the logarithm of 19 for the base 2. The problem then becomes: Given an integer, find the exponent of 2 that is going to lead to that integer—in other words, find the logarithm base 2. Then, if you know the logarithm, if you know the exponent, what is the number that comes out of that? This introduces the idea of the exponential function. We're interested not just in 2^2 and 2^3 and the integer powers of 2 or the simple fractional powers of 2. We're interested in 2 raised to a power that could be any real number. We're now interested in the exponential as a function.

Now in order to be able to do the kinds of calculations that Napier needed to do—and that the astronomers, people like Kepler, needed to do—if we want 7-digit accuracy, then what we're going to need to be able to do is find tables of logarithms for all of the numbers in increments of 1 in 10 million. I can simplify the problem slightly by just asking for the logarithms of the numbers between 1 and 10. If I know the logarithms of the numbers between 1 and 10, and I've got two numbers like 19 and 33, I can divide 19 by 10, and 33 by 10, and I'm taking the product of 1.9×3.3. If I know that product, I take that answer, and I just put the factors of 10 back in. So, I'll take my answer and multiply it by 100.

But there still is the problem of how to get the logarithms of the numbers from 1 to 10 down to the kind of accuracy that you need. One of Napier's great insights is that if you know the logarithms for base 2, you can use them to find the logarithms for base 3 or any other base you want to use—that there is always going to be a simple conversion factor—so that if I calculate the logarithms for base 2, I can use them to find logarithms for base 3, or base 5, or base 10, or whatever base I might want to use.

Now a base that actually makes sense if you're going to do 7-digit accuracy is the base 1.0000001—so that is $1 + 1/10{,}000{,}000$. If I can find the logarithms for that base, and the powers of that base are fairly easy to calculate—if I want to multiply a number by 1.0000001, I just take that number, and that same number with the decimal point moved over seven places, and add those two together. If I want to take those powers, that will give me the 7-digit accuracy that I need. But in order to get the logarithms for all the numbers from 1 to 10, I'm going to need 23 million powers of 1.0000001.

Now that's an impossible task, and Napier realized that he couldn't do those 23 million calculations. But his insight (that once you know the logarithms for one base, there will always be a simple conversion factor that enables you to find the logarithms for any other base) actually enabled him to find these 23 million values very easily. What he actually did was to first find the powers of 1.01 that go from 1 up to 10. You only need 230 powers in order to accomplish that. Then what he did was to interpolate between those. He then took the powers of 1.0005, and 20 powers of that enabled him to get numbers between 1 and 1.01, between 1.01 and 1.02, and so on.

Then he looked at the first 50 powers of 1 + 1/100,000, and that enabled him to interpolate between the values that he already had. Then he looked at the first 100 powers of 1 + 1/10,000,000, and that enabled him to interpolate beyond the remaining powers. So, with just a few hundred calculations, he was actually able to set up tables that would enable somebody to find logarithms—logarithms to the base 1.000001—for all numbers between 1 and 10 to the desired 7-digit accuracy. But he also realized that since it was so easy to change these bases back and forth, there was no reason to stick with the base 1.0000001. He could choose any base he wanted.

The question that he next grappled with is: What base to use? What he did was to consider two lines with points moving along each line. On the first line, we've got the point that we want to take the logarithm of. So, we're going to take the logarithm of a real number, and that real number is traveling along this first line. On the second line, we look at the image of that point under the logarithm. So, I've got a point like 3. On the second line, the position is going to be at the logarithm of 3. When the first point gets to 5, the point on the second line is moved up to the logarithm of 5. Get up to 13, and it's moved up to the logarithm of 13. The point on the first line travels at a uniform speed.

How fast does the point on the second line move? Well, what Napier realized is that the point on the second line moves at a speed that is inversely proportional to the distance that the point on the first line is moved. In other words, if the first point has moved to position x, the speed of the point on the second line is going to be some constant divided by x. The value of that constant is going to depend on the base that I am using for the logarithm. What Napier decided was to choose the base that would give him a speed where that constant of proportionality was exactly equal to 1.

Now for those who know a little bit about calculus, you may have realized at this point that that's exactly equivalent to finding the slope of the derivative of the logarithm function, and that the slope of the derivative of the logarithmic function is inversely proportional to the value at which you're evaluating it, and that it is this particular logarithm that Napier chose that is what today we call the *natural logarithm*. Napier never actually worked out what that base needs to be, but it was later observed that it's a number between 2 and 3; it's about 2.71. It's a number that would later be referred to by the letter

e, and e now is known as the *base of the natural logarithm.* It is the base of Napier's logarithm. The actual name "natural logarithm" would be invented later on in the 1600s.

While Napier used this particular base, 2.71, Henry Briggs down in London got a hold of Napier's work on logarithms and recognized that a lot of the calculations would be much easier if instead of using this strange number 2.71, Napier would actually go back and use the base 10. If you're actually taking powers of numbers, a base 10 makes it much easier to work with. So, Briggs went up to Edinburgh in 1615, and he convinced Napier, instead of working out his table of values with 2.71 as the base, to use 10 as the base. Briggs would then go on to develop his own tables of logarithms, and in 1628, Briggs would actually publish a table of logarithms that was accurate to 10-digit accuracy.

But, Napier's logarithm, this 2.71 base, would wind up having a very important role to play in the calculus. In the mid-17th century, two Belgian Jesuits, Gregory of St. Vincent and Alfonso Antonio de Sarasa, would consider the problem of finding the area underneath the curve of a hyperbola, so that's a curve that's simply the reciprocal of the variable, so the graph of $y = 1/x$. What they were able to show is that if you look at the area underneath this curve—starting at $x = 1$, going up to some value, say t—the area underneath the hyperbola from 1 to t is actually given by Napier's logarithm of t. It's given by this natural logarithm of t. So, we get both the exponential function and the logarithmic function coming in, and they would be very important concepts as mathematicians then went on to develop the calculus. I'll talk about that more in later lectures.

For the next lecture, I want to go back to the beginning of the 1600s and take a look at another one of the important influences of astronomy, and that is on the work of Galileo and the beginning of the understanding of the mathematics of motion.

Lecture Eleven

Galileo and the Mathematics of Motion

Scope:

This lecture explores the work of Galileo Galilei, an acknowledged great thinker, but one whose precise contributions are difficult to pin down. Much of his work was foreshadowed by earlier scientists, and much of it had to be revised or completed by later scientists. Galileo's greatness can be located in his ability to ask the right questions and his understanding that the answers could be found in mathematics. Galileo believed that the Earth travels around the Sun and sought to explain why, if that was the case, people on Earth don't sense the planet's motion. To this end, he studied gravity and the laws of motion. He explored velocities, partially worked out the concept of inertia, and anticipated the method of finding total distance traveled by calculating the area under a curve that represents velocity. Galileo's reliance on mathematical models to explore physical phenomena would lead to further discoveries in science and the development of analytic geometry and calculus.

Outline

I. In Lecture One, we mentioned that five strands of mathematics would come together in the 17th century: algebra, geometry, astronomy or astrology, mechanics, and the mathematics of motion. The first figure in which we see this fusion is the great scientist Galileo Galilei (1564–1642).

  A. Galileo is universally acknowledged to be an important scientist, but as E. J. Dijksterhuis, one of the great historians of science in the 20th century, said of him, "On the question of what precisely his contribution was, and wherein his greatness essentially lay, there seems to be no unanimity at all."

  B. Virtually everything Galileo did was foreshadowed—in some cases, fully realized—in the work of earlier scientists, and much of his work had to be revised or completed by later scientists. His greatness lay in his talent for asking the right questions and knowing where to look for solutions.

C. In particular, Galileo knew that the key to understanding the motions of the planets (celestial mechanics) could be found in mathematical modeling. This emphasis on mathematics set Galileo apart from other thinkers of his day, and his work would eventually lead to the development of calculus.

II. Galileo was trained as an algebraist. He got his first job in 1585 in Florence and would hold a succession of positions in Sienna, Vallombrosa, and Pisa. In 1592, he was appointed to a position in Padua, where he may have met John Napier in 1594. In 1610, he published *The Starry Messenger*.

A. Some years earlier, Galileo had gotten hold of one of the first telescopes, originally built in the Netherlands. He used it to see not only objects that were far away on Earth but also the Moon and the other planets in the heavens. He was the first to observe moons orbiting other planets and geographical features on the face of the Moon.

B. *The Starry Messenger* was a popular book that laid out Galileo's arguments for the fact that the Earth orbits the Sun rather than the other way around. He also made this point quite clearly in a later book, *Letter to the Grand Duchess* (1616).

C. After the publication of *Letter*, Galileo was brought before the Inquisition in Rome and warned not to state categorically that the Earth traveled around the Sun. He could use that idea as a basis for a model, but he was not to assert it as fact because it contradicted passages in the Old Testament in which the Earth was described as stationary.

D. Galileo was convinced that the Earth in fact moved. To support that view, he knew he had to explain why people on Earth were not aware of the planet's motion. Thus, he began to try to model gravity, inertia, and other aspects of physics that would explain how we could live on a moving Earth without being aware of its motion.

E. Galileo's book explaining his understanding of the universe appeared in 1632: *The Dialogue Concerning the Two Chief Systems of the World*.

1. The book is written as a debate among three protagonists concerning the issue of whether the Earth travels around the Sun or vice versa.

2. The three protagonists are Salviati, Sagredo, and Simplicio. Salviati is an intelligent layman who is trying to learn the truth, Sagredo argues that the Earth travels around the Sun, and Simplicio holds to the belief that the Earth is stationary.
3. Although the book claimed to be impartial, it clearly argued in favor of the idea that the Earth orbited the Sun. For this reason, Galileo was again called before the Inquisition in Rome.
4. Galileo recanted and was forbidden from publishing *Two Chief Systems of the World.* He was restricted to house arrest at his home in Arcetri for the rest of his life.

III. Galileo realized that the key to understanding why we don't sense the motion of the Earth is gravity.

A. To Aristotle and later scientists/philosophers, gravity (*gravitas*) was a property inherent in certain bodies—the tendency of certain objects to move toward the center of the universe.

1. The Earth was at the center of the universe because it had more gravity than anything else.
2. As opposed to gravity, an object could have *levity*—a tendency to move away from the center of the universe. Fire rises because it has levity.

B. We can see one of the problems with this theory of gravity and levity in the act of throwing a ball. When a ball is thrown, it first moves away from Earth. Why doesn't it immediately start moving toward the center of the universe?

1. One explanation that was offered was the *theory of impetus*—the idea that a moving object accumulates an impetus, or tendency to keep moving in a given direction.
2. Medieval philosophers were able to explain the motion of a thrown object by postulating that the object acquires a certain impetus from the hand; they expected the impetus to gradually dissipate as the object traveled through the air and finally fell to the ground.
3. Early explanations of the motion of cannonballs used this idea. We have illustrations showing a cannonball being fired from a cannon, traveling in a straight line

until it runs out of impetus, and finally falling vertically to the ground. Clearly, however, a cannonball travels along an arc.

C. One of the properties of gravity tested by Galileo was the idea that a heavy object falls faster than a lighter object.

1. We often hear that Galileo went to the top of a tower, dropped two balls of unequal weight, and observed that they hit the ground at the same time. We have no evidence that Galileo ever conducted this experiment, but it was conducted by the Belgian-Dutch scientist Simon Stevin (1548–1620) in 1586.

2. Stevin is an important figure for a number of reasons. In 1585, he wrote *La Thiende*, a book advocating the adoption of the Islamic system of decimal fractions—tenths, hundredths, thousandths, and so on—over the Babylonian system that used sixtieths. At the time, the two systems existed side by side, with merchants using decimal fractions and scientists using sixtieths.

3. Thomas Jefferson had an English translation of Stevin's book and was so impressed by his argument that he advocated the use of the decimal system in American money.

D. Galileo is also credited with a thought experiment related to falling objects: What if we drop two balls connected by some kind of bar? They now weigh at least twice as much and should fall twice as fast. What if the bar is very thin? Do the two balls constitute one object? What if a piece of string or a thin thread connects the two balls? Does that make them one object or two?

1. This thought experiment illustrates that the rate at which something falls cannot be dependent on its weight.

2. This idea is often credited to Galileo, but in fact it goes back to the 16th century and the work of Giovanni Benedetti.

IV. Another question Galileo explored was whether the velocity of a falling object—which increases as the object falls—increases as a function of time or a function of distance.

A. Velocity can be expressed either as a function of time or as a function of distance, but which expression is easier? Scientists up until the early 1600s debated this question.

B. Galileo eventually concluded that velocity increases at the same rate over each interval of time. We know that Galileo conducted experiments with balls rolling down inclined planes, showing that in each unit of time, the ball picks up the same increase in velocity. Thus, velocity is most easily expressed as a function of time.

C. With the understanding that velocity increases by the same amount for each unit of time, we might ask: How far does a dropped ball fall over a given period of time? To answer this question, Galileo represented the velocity at each interval of time by a small vertical line, a representation that would serve as an important precursor to calculus.

   1. If a ball is dropped, starting with velocity zero, and we align the vertical lines that designate the successive velocities, they form a triangle.
   2. Galileo realized that the distance the object would travel is precisely the area of this triangle: the base (the time that has elapsed) times half the height (the final velocity). In other words, the distance traveled under uniform acceleration is the elapsed time multiplied by half of the final velocity.
   3. The idea of representing the velocities as vertical lines was a precursor to the development of analytic geometry.

V. Galileo also explored the vector decomposition of velocities.

A. Think again of throwing a ball. Galileo broke the velocity of the ball down into two parts, a horizontal component and a vertical component. He assumed that the horizontal component of the velocity does not change much, but the vertical component does.

B. We can treat the vertical component of the velocity as if we're working with a falling object. The vertical motion is initially upward or positive, but it decreases by the same amount in each unit of time.

C. Using this idea of decomposition, Galileo was the first person to show that the trajectory of a thrown object must be a parabola.

D. Galileo's idea of decomposing a velocity into two orthogonal components (two parts at right angles to each other) was foreshadowed by the work of Simon Stevin in decomposing forces. Stevin wrote a work called *Elements of the Art of Weighing*, in which he studied objects on an inclined plane.

1. Imagine that we have a slanted plane (with no friction on its surface) with an object sitting on it. The force of gravity is pulling this object down. How much force do we need to apply in order to keep the object stationary on the plane?
2. The actual force acting on the object is gravity, which is straight down, but the object can only move along the slanted plane.
3. Stevin realized that the amount of force needed to counteract the tendency of the object to slide could be determined by decomposing the vertical force into two parts, one force that ran parallel to the plane and another force at right angles to the plane.
4. We construct a rectangle so that one side is parallel to the plane and the vertical force (due to gravity) is the diagonal of the rectangle. The length of the side of the rectangle that is parallel to the plane describes the force needed to keep the object stationary.

VI. Galileo looked into the question of inertia, which he knew was also involved in explaining why we don't sense the motion of the Earth.

A. Initially, inertia was closely connected to the idea of impetus; it was thought to be a tendency of bodies to remain at rest. To start motion and continue motion, inertia must be overcome.

B. Galileo was the first scientist to see that inertia is a tendency of an object to keep traveling in the direction it has been traveling. If an object is in motion, it tends to stay in motion unless a force acts on it. If an object is stationary, it tends to remain stationary unless a force acts on it.

C. Galileo tried to use the concept of inertia to explain why we have no sense of movement even though we stand on an Earth that is spinning at an incredible speed.

   1. Galileo considered "circular inertia"—the idea that if we are moving in a circle, we tend to continue this circular motion, and that is why we don't sense the motion of the Earth.
   2. Later scientists would realize that circular inertia does not exist. Inertia only operates along straight lines.

D. It has been suggested that Galileo was not able to understand the tendency of objects to keep moving in a straight line at the same velocity forever because that idea assumes that space goes on forever, and Galileo was not prepared to accept the fact that space is infinite.

   1. He still believed that the universe was enclosed within a sphere, and in fact, circles were important to his understanding of the universe.
   2. Galileo's real contribution, however, was the recognition that an understanding of the world required mathematics. The universe, he said, "is written in the language of mathematics, and its characters are triangles, circles, and other geometric figures without which it is humanly impossible to understand a single word of it."

VII. In our next lecture, we will continue with Pierre de Fermat and René Descartes, who transformed Galileo's idea of representing velocity as a sequence of vertical lines into analytic geometry and thus set the stage for the development of calculus.

Suggested Readings:

Cohen, *The Birth of a New Physics*, chaps. 4, 5.

Dijksterhuis, *The Mechanization of the World Picture*, 324–85.

Gindikin, *Tales of Mathematicians and Physicists*, 27–78.

Katz, *A History of Mathematics*, 314–21, 420–25.

Naess, *Galileo Galilei*.

Questions to Consider:

1. Galileo is universally considered to be an important figure in the development of science, but there is little agreement when it

comes to pinning down his innovations. What do you see as his most significant contributions?

2. What *was* Galileo's contribution to our understanding of gravity?

Lecture Eleven—Transcript

Galileo and the Mathematics of Motion

In my very first lecture I talked about the 17th century as a pivot point, and I also talked about the five strands that would come together in the 17th century: algebra, geometry, astronomy or astrology, mechanics, and the mathematics of motion. The first figure in which we really see these five strands coming together is the great scientist Galileo Galilei, born in 1564; he would die in 1642. He is a controversial figure insofar as knowing what it was that he really accomplished. There is a quote that I really love from E. J. Dijksterhuis. Dijksterhuis was one of the great historians of science in the 20th century, and he wrote about Galileo:

> The whole history of science perhaps cannot point to a single figure about whom opinions differ so widely as about Galileo. No one, indeed, is prepared to challenge his scientific greatness. … But on the question of what precisely his contribution was and wherein his greatness essentially lay there seems to be no unanimity at all.

In fact, as we look at what Galileo accomplished, we see that virtually everything that he did we can find foreshadowed in the work of earlier scientists—and often not merely foreshadowed, but often quite fully worked out. Many of the important contributions that he made, he never got quite right. It would fall to later scientists in order to complete them. But for myself, where I see Galileo's greatness is that he knew the right questions to ask. He knew what was important, and he had a good sense of where to look for the solutions. Most particularly, he realized that if we wanted to understand the motions of the planets, if we wanted to understand celestial mechanics, we really needed to look at mathematical modeling. It's this very important emphasis on the mathematics that really sets Galileo apart from others. He didn't work out the techniques, but he showed us how to approach the problems, and his work then would lead on into the eventual development of the calculus.

Galileo was really trained as an algebraist, as a mathematician. He got his first job in 1585 in Florence. He would then hold a succession of positions in Sienna, in Vallombrosa, in Pisa—and in 1592, he was appointed to a position in Padua. This is where he would have been in 1594, when John Napier visited Italy. We know that Napier probably got to Padua, which makes it very likely that the two of them actually met each other. In 1610, Galileo would return to Pisa. There was a very important book that he published in 1610 called *The Starry Messenger*. Some years earlier, Galileo had gotten ahold of one of the first telescopes. This originally had been built in the Netherlands, in the lowlands of Belgium and Holland.

What Galileo did was to use it not just in order to see objects that are far away on the Earth, but actually to look at the heavens, and to look at the Moon, and to look at the planets—and to observe what was out there, and to observe the first moons of other planets that were known to exist, thus showing that not everything went around the Earth, because there were moons that go around the other planets, and to see things like the mountains on the Moon—thus showing that the Moon was not ethereal and perfect—that the Moon, in fact, had imperfections on it. *The Starry Messenger* of 1610 was a popular book that he wrote explaining what he understood about the heavens and the arguments for the fact that it is the Earth that travels around the Sun, rather than the Sun that travels around the Earth. This was made quite explicit in a book published in 1616, *Letter to the Grand Duchess*, in which he lays out this point of view quite clearly.

This was the first time that he was pulled before the Inquisition in Rome, and he was warned at this point that he should not state that he knew categorically that the Earth traveled around the Sun. It was fine for him to put that out as a possible basis for a model, but he was not allowed to say that this in fact is what was happening, because there are passages in the Old Testament—especially in the Psalms—that talk about the Earth being on a firm foundation and the Earth never moving. So, the Catholic Church believed that it was important not to state that, in fact, the Earth could move.

But Galileo was convinced that in fact the Earth moved, and he realized that if he was going to support that point of view, he had to be able to explain how it was possible for the Earth to move and for us not to be aware of its motion. So, he began trying to model gravity, and inertia, and all of the physics that would go behind explaining how we could be on a moving Earth without being aware of its motion. The great book that he wrote that explains his physical understanding of the universe would finally appear in 1632, called *The Dialogue Concerning the Two Chief Systems of the World*. He was very careful, or he thought he was being very careful, not to put this out—the idea that the Earth is moving around the Sun—as something that he actually believed. He wrote the book with three protagonists who were debating this issue of whether the Earth goes around the Sun, or it is the Sun that goes around the Earth.

His three protagonists were Salviati, Sagredo, and Simplicio—one of whom was an intelligent layman; one of whom was a person who believed that in fact it is the Earth that goes around the Sun; and one who argued that in fact the Earth is stationary and the Sun goes around the Earth. The person who held to the position that it is the Earth that is stationary was given the name Simplicio, which perhaps suggests that Galileo's sympathies were not with him. He is somewhat portrayed as the simpleton of the group. Certainly, the Inquisition was very much aware of the fact that while this book claimed to be impartial, there was a very clear argument in favor of the Earth traveling around the Sun. For this reason, Galileo was called before the Inquisition in Rome. Many of us know the story, how he recanted. He agreed that it must be the Earth that is stationary, and he was forbidden from publishing this work, *The Dialogue Concerning the Two Chief Systems of the World*. He was restrained, restricted to house arrest at his home in Arcetri for the rest of his life.

But I want to look at the mathematics that he did, and especially the mathematics that appears in *The Dialogue*. One of the things that Galileo realized is that the key to understanding what is happening, why we don't sense the motion of the Earth, has got to be gravity. To understand gravity, we must go back to the Aristotelian viewpoint of what gravity is and what Galileo was changing about our understanding of gravity.

To Aristotle and the later scientists/philosophers, gravity (*gravitas*) was a property inherent in certain bodies. It's a tendency of certain objects to move toward the center of the universe. If the Earth is at the center of the universe, that is because it has more gravity than anything else. If you drop an object—if I take a book and let go of it—and it falls, that is because the book has gravity. It is trying to get to the center of the universe. As opposed to gravity, an object can have *levity*—which is a tendency to move away from the center of the Earth. Fire rises because it has levity.

One of the problems that you encounter as you try to explain what is happening in terms of gravity and levity is the problem of what happens when you throw an object. If I take a ball and I throw it, as long as the ball is in my hand—because it's in contact with my hand—it can move away from the center of the Earth. But as soon as it leaves my hand, it continues moving away from the center of the Earth. I can throw it up in the air, and it continues moving away from the Earth. Why doesn't it immediately start moving toward the center of the universe? The explanation that eventually would be found is the *theory of impetus*—the idea that as I'm moving this object, it begins to accumulate an impetus, a tendency to keep moving in a given direction.

So, to the medieval philosophers, they were able to explain this motion of a thrown object as something that acquires a certain impetus from the hand, and they expected that this impetus would gradually dissipate as the object traveled through the air—until finally the impetus ran out, and the object would then fall to the ground. Early explanations of the motion of cannonballs used this idea. People drew illustrations of the motions of cannonballs that showed a cannonball being fired straight out of the cannon, traveling in a straight line until it ran out of impetus, and then falling vertically down to the ground. Hopefully, it was above the target at the moment that it began falling—but it was pretty clear that this is not really the way in which a cannonball travels. A cannonball travels along an arc. So trying to explain that arc and what is going on with this gravity, with this tendency to move toward the center of the Earth, was one of the important problems not just for Galileo but also for the scientists of the 16th century.

As Galileo tested the ideas behind gravity, one of the ideas that he tested is this idea that if you've got a heavy object, it's going to fall faster than a lighter object. There is a common story told that Galileo went up to a tall tower—perhaps the Leaning Tower of Pisa—and took two balls of unequal weight and dropped them and observed that they hit the ground at the same time. There actually is absolutely no evidence that Galileo ever conducted this experiment, but we do know somebody who did—the Belgian, and then later Dutch, scientist Simon Stevin, who lived 1548 to 1620. It is known that in 1586, he went up to the top of a church tower and actually performed this experiment of dropping two balls and observing that even though they have very unequal weights, they both hit the ground essentially at the same time.

Stevin is an important figure for a number of reasons. There are a number of things that Galileo did that were prefigured in Stevin, and I'll come back to him a little later this lecture. But I just want to mention a very influential book that he wrote in 1585 called *La Thiende*, which advocates for the adoption of decimal fractions. If you'll recall when I was talking about the Islamic mathematicians, I said that they adopted the idea of decimal fractions—tenths and hundredths and thousandths—using the decimal system. They also continued to use the system that had been inherited from the Babylonians that used sixtieths, and sixtieths of sixtieths, as we get now because an hour is divided into 60 minutes, and a minute is divided into 60 seconds. What happened in the Islamic world and in the early Italian world was two systems that existed side by side. For most of the merchants, they found tenths and hundredths (decimal fractions) to be very easy to work with. In scientific work—especially astronomical work, but also more general scientific work—they would use sixtieths, and sixtieths of sixtieths.

It was Simon Stevin who argued that you really want to use the decimal fractions—tenths, hundredths, thousandths, and so on—and he wrote this little book to argue for this point. We happen to know that Thomas Jefferson had an English translation of Simon Stevin's book and that Jefferson was so impressed by this argument that when the new United States government set up its own money (the dollar), one of the questions that came up is: How should we subdivide the dollar? The British system, of course, was to divide a pound into 240 pence. Rather than dividing a dollar into 240 pennies, Jefferson took this work of Simon Stevin and argued that we should divide the dollar up into 100 pennies—that we should work with the decimal system for our dollar. So, Simon Stevin had a direct influence on American currency. Simon Stevin also wrote very influential books on arithmetic and on algebra.

There is another experiment—this time a thought experiment—that Galileo is credited with doing on falling objects—again, with whether two objects of different weights should hit at the same time or whether the heavier object should hit first. That involves the question: What if I take two balls and I connect them by some kind of a bar? Now suddenly they weigh at least twice as much, and they should fall twice as fast—but what if that's a very thin bar? Is that still one object? What if instead of being a bar, it's simply a piece of string that connects the two balls? Does that now make them one object? What if it's just the lightest gossamer thread that connects the two balls? Does that now make them one object? Obviously, once you get a very thin connection, you need to think of them as two objects. This whole question raises the realization that, in fact, the rate at which something falls cannot be dependent on its weight. That was something that is often credited to Galileo. In fact, that goes back also into the 16th century to Giovanni Benedetti, who also studied this question of falling objects.

Galileo looked at this question of falling objects, and one of the questions that was still out there, that was still undecided, is whether the velocity—which increases as an object falls—is increasing as a function of time or as a function of distance. Actually, you can express the velocity either as a function of time or as a function of distance. The real question is: Is it easier to express it as a function of distance or a function of time?

The scientists up until the early 1600s had debated this. We see in Galileo's own writings that he himself went back and forth on this, taking different points of view and eventually coming to the conclusion that you've got to look at the velocity as increasing at the same rate over each interval of time. In fact, we do know that Galileo conducted experiments with balls rolling down inclined planes that showed that in each unit of time the ball picks up the same increase in velocity. So, velocity is most easily expressed as a function of time, and this would be a very important insight.

That now raises the question: If you've got an object that is falling, such as a ball falling under the influence of gravity, we let go of that ball, and its velocity is increasing. So, for each unit of time, the velocity increases by the same amount. How far does the ball fall over a given period of time? Galileo did something very important here that would be a precursor to calculus, to integration, and that was to represent the velocity at each piece of time by a small line. So, we would take the velocity as initially zero, and then a short line that would represent the velocity a short time later, and then a slightly longer line that would represent the velocity slightly later, and a slightly longer line that would represent the velocity later. You put all of these lines together, and you get a triangle.

What Galileo realized was that the distance that the object would travel is precisely the area of this triangle. It can be represented by the area of this triangle, and the area of the triangle is the base (which is the time that has elapsed) times half of the height. You take the final velocity and take half of that, so that the distance traveled under uniform acceleration—the velocity changes by the same amount in each unit of time—the total distance traveled is the amount of time it takes multiplied by half of the final velocity. This idea of representing the velocity as little vertical lines would really be the key leading into the development of analytic geometry that I will be talking about in the next lecture.

Something else that Galileo did that would be extremely important is the idea of the vector decomposition of velocities. I'm going to take a ball, and I'm going to throw it. What Galileo did was to take the velocity of the ball and break it down into two parts—a horizontal component and a vertical component. What Galileo assumed is that the horizontal component is not going to change—it's certainly not going to change very much—but the vertical component of the velocity is going to change. We can treat the vertical component of the velocity as if we're working with a falling object. So what we can do is to take this ball that is being thrown and decompose the motion into a horizontal motion at a constant velocity, plus a vertical motion that initially is going up and then is coming back down. The vertical component is decreasing by exactly the same amount in each unit of time.

By doing this kind of decomposition of the velocities, the horizontal component does not change; the vertical component decreases by the same amount in each unit of time. Galileo was able to show that the path is going to have to be a parabola, and this is the first time we get the description of the trajectory of a thrown object, or the trajectory of a cannonball, being shown to be modeled by a parabola.

This idea of decomposing a velocity into two orthogonal components (two pieces that are at right angles to each other) had some precursors in ideas of Simon Stevin. Stevin was not decomposing velocities. What Stevin did was to consider decomposing forces, and one of his important works was *Elements of the Art of Weighing*. One of the things that he looked at there was the problem of an object that is sitting on an inclined plane.

I've got a slanted plane, and I've got an object sitting on it, and I've got the force of gravity that is pulling this object down. We assume that there is no friction on this surface. The question is: How much force do we need to apply in order to keep that object stationary on the plane? The actual force that is acting on this object is a force that is straight down, but the object would have to move parallel to the surface of this inclined plane. What Simon Stevin realized is you could figure out the amount of force that was needed to counteract the tendency of the object to slide by taking that vertical force and decomposing that into two pieces: one force that ran parallel to the plane and another force that was at right angles to the plane. You simply construct a rectangle so that the vertical force (due to gravity) becomes the diagonal of a rectangle. The side of the rectangle that is parallel to the plane—that is the component of the force that is pushing this object down the plane. That is the force that is needed in order to keep the object stationary. So, this idea of taking a force and decomposing it is something that we get from Simon Stevin. Galileo took Stevin's idea, and he then applied it to a decomposition of velocities, which turned out to be a very useful way of understanding trajectories.

Something else that Galileo looked at was the question of inertia. As Galileo realized, inertia really would be the key to understanding why we don't have this sense of being on a moving Earth, of moving around the Sun at such incredible speeds. Now inertia initially was very closely connected to this idea of impetus, this belief that inertia is something that is a tendency for bodies to remain at rest. To get something going and to keep it going, you've got to overcome inertia. Galileo was the first scientist to really realize that inertia is a tendency to keep going the way you have been going. So, if you're in motion, you're going to tend to stay in motion unless there is a force that acts on you. If you are stationary, you are going to tend to remain stationary unless there is a force that is acting on you.

But Galileo didn't quite get inertia correct. He was trying to explain why we're able to stand on an Earth that is spinning at such an incredible speed without any sense that we are spinning on this Earth. He tried to explain this in terms of inertia. He believed in a "circular inertia"—that if you're moving in a circular pattern, you're going to tend to stay in a circular motion, and that is why we don't sense this motion of the Earth—that we have this circular inertia that is tending to take us around the Earth, and also that we have this circular inertia that is moving us around the Sun, and we're not aware of that motion. Well, he was wrong. As would later be discovered, there is no such thing as a circular inertia; there is only an inertia in straight lines.

It has been suggested that one of the reasons that Galileo was not able to realize this idea of a tendency to keep moving in a straight line at the same velocity forever is because that really assumes that space goes on forever. Galileo was not prepared to accept space that was infinite. To him, he still believed in a universe that was enclosed within a sphere, and in fact circles were extremely important to his understanding of the workings of the universe. Galileo was someone who never accepted Kepler's law that, in fact, the planets moved in ellipses. Galileo always held to his belief that somehow everything really was moving in circles.

But as I said at the beginning of the lecture, and I want to come back to now, Galileo's real insight, his real contribution, was the realization that if you want to understand the world around us, you need to use mathematics. You need to look at mathematical modeling. I'd like to close with a quote from Galileo that really says this very clearly and would really set the stage for the mathematics that would follow in the 17th century. He wrote:

> Philosophy is written in this grand book, the universe, which stands continually open to our gaze. But the book cannot be understood unless one first learns to comprehend the language and read the characters in which it is written. It is written in the language of mathematics, and its characters are triangles, circles, and other geometric figures without which it is humanly impossible to understand a single word of it; without these one is wandering in a dark labyrinth.

The key to understanding the universe is to know the language of mathematics, and Galileo then would set the stage for this. In our next lecture, we will continue with Fermat and Descartes, taking Galileo's idea of representing velocity as a sequence of vertical lines and turning that into analytic geometry—and so setting the stage for the development of calculus.

Lecture Twelve

Fermat, Descartes, and Analytic Geometry

Scope:

Pierre de Fermat, a lawyer from Toulouse, received his mathematical training in the algebraic tradition of François Viète. He read Latin translations of classical Greek works in mathematics and applied modern algebraic techniques to push this work forward. One of his earliest insights was a method for translating geometric problems into algebraic equations and vice versa, a discovery he shared with René Descartes. Fermat is the inventor of differential calculus, creating it specifically to solve the problem of finding the maximum value of a function. He discovered the rules for finding the area under the graph of $y = x^n$ for an arbitrary integer n. Fermat also did work with perfect numbers and Pythagorean triples, leaving his last theorem as a legacy to be solved by 20th-century mathematicians.

Outline

I. In this lecture, we explore the work of Pierre de Fermat (1601–1665) and René Descartes (1596–1650). We begin with Fermat, who was instrumental in the development of analytic geometry and calculus.

   A. Fermat was born near Toulouse and studied mathematics in Bordeaux with some of the disciples of François Viète, a French algebraist from the 1500s. Fermat then went on to Orléans, where he earned a law degree before returning to Toulouse. He secured a position as a counselor to the Parliament and rose quickly in government service.

   B. Fermat viewed mathematics as an avocation and published little during his lifetime. His son published his manuscripts after his death.

   C. Fermat's work with mathematics began in the 1620s, when he learned of a lost book by Apollonius of Perga, the author of the *Conics*. This lost work, *Plane Loci*, dealt with two-dimensional curves. Fermat took it upon himself to prove some of Apollonius's results.

   D. Fermat invented an entirely new way of approaching geometric problems that is now called *analytic geometry*. He interpreted geometric statements algebraically, though the correspondence can also be used to create geometric representations of algebraic expressions.

1. When we graph an algebraic expression, we use a horizontal axis and a vertical axis. We then plot points for each pair of values (x, y) that satisfies the equation.
2. As we plot the points on the *Cartesian coordinate axis*, the algebraic expression is transformed into a geometric curve.
3. Fermat translated the geometric statements of Apollonius into algebraic statements; he then used advanced algebra to find simple proofs of the geometric theorems.

E. Fermat constructed his graphs without a vertical axis. He represented one of the variables on a horizontal axis and then simply marked off the distance to the other variable, similar to what Galileo had done in representing velocity—but Fermat used points rather than Galileo's vertical lines.

F. In 1637, Fermat published his idea in *Introduction to Plane and Solid Loci*.

1. In the same year, René Descartes independently came up with the same idea for translating between geometric curves and algebraic expressions. Descartes published his work, *Geometry*, in a more comprehensive treatise called *Discourse on the Method for Rightly Directing One's Reason and Searching for Truth in the Sciences*.
2. Descartes' explanation of how to translate between geometric and algebraic expressions became much more popular than Fermat's, partly because Descartes used more efficient algebraic notation.

II. Descartes and Fermat differed in their approach to moving between geometry and algebra.

A. Descartes, like Fermat, started with the idea of interpreting a geometric object algebraically and then looked for geometric conclusions by studying the algebra.

B. Fermat, however, realized that he could get powerful results by moving in the other direction. He could graph an algebraic equation, such as an equation that describes a quadratic polynomial, and work with the resulting parabola instead of the algebraic expression—perhaps seeing ideas that were not readily apparent in looking at the algebra alone.

1. Think, for example, of a parabola that opens downward. Looking at this geometrically, we can pose the question: What is the area between the horizontal axis and the parabola?
2. Again, looking at an expression geometrically, we can also see where an algebraic equation hits its largest or smallest value within some range of values.
3. The ability to translate an algebraic expression into a geometric representation led directly to the development of much of calculus.

III. As mentioned in an earlier lecture, Descartes coined the term *imaginary number* for the square root of a negative number.

A. In the 17th century, mathematicians were still unsure about the legitimacy of negative numbers. Earlier Chinese, Indian, and Islamic mathematicians had been divided on this issue.

1. Descartes was one of the last western European scientists to consider negative numbers to be less legitimate than positive numbers. He viewed the negative root of a polynomial (a negative solution to a polynomial equation) as a "false root."
2. By the middle of the 17th century, negative numbers were accepted as fully legitimate solutions to problems.

B. Descartes' work *Principles of Philosophy* (1644) set science back in some ways.

1. In four volumes, Descartes sought to lay out the principles on which all knowledge of the universe could be based. His foundations would replace those originally established by Aristotle.
2. Descartes stated quite forcefully that there could be no action without an actor, no action at a distance. If something is moving, it is doing so because something else is moving it.
3. He explained the motion of the planets by postulating a swirling ether that pervaded the universe and moved the planets with its own motion. People on Earth don't sense

the motion because we are also embedded in the ether and moving with it.

4. When Isaac Newton published his *Principia* in the later 17th century, he was unable to explain how gravity could act at a distance without an intermediary. Many people criticized Newton because he could not come up with a Cartesian-type mechanism to explain the action of gravity.

IV. Although Fermat published little, he was an active correspondent in mathematics; in particular, he exchanged letters with Father Marin Mersenne (1588–1648), a Catholic priest in Paris.

A. Mersenne was a key figure in the exchange of information among scientists of the time. He was a great friend of Descartes' and Christiaan Huygens's and a supporter of Galileo.

B. In 1636, Fermat tackled the problem of finding the area underneath a curve described by an algebraic expression, using what would come to be understood as integral calculus. Fermat used the method of exhaustion discovered by Eudoxus of Cnidus. In the same year, another mathematician, Gilles de Roberval, found the same formula as Fermat for the area under a curve that corresponds to an arbitrary polynomial.

C. In 1639, Fermat solved an important problem: Given an algebraic expression, how can its maximum (its greatest value) or its minimum (its least value) be found?

1. The idea is to represent the algebraic expression as a curve and then look at the line that is tangent to the curve at each point.
2. The slope of the tangent will be positive if the expression is increasing and negative if the expression is decreasing. For an expression that is at its highest or lowest value, the tangent line will have a slope of zero.
3. Fermat then looked at the problem of determining the slope of the tangent line at a single point. He considered two points that are very close to the point in question on the curve. It's possible to construct the equation of the straight line that goes through those two points, and the

slope of that line is the rise over the run, or the change in the y value divided by the change in the x value.

4. Fermat saw that as the change in the x value gets smaller, the change in the y value also gets smaller. This ratio, in which both the numerator and denominator get closer to zero, will approach the slope of the tangent line at that particular point.
5. When we want to find the maximum or minimum value, we follow this procedure and look for those points where the slope of the tangent line is exactly equal to zero.
6. The fact that the slope of the tangent line is zero does not guarantee that we have found a maximum or a minimum value. But for most algebraic expressions, there are very few places where the function has a horizontal tangent line; thus, the number of places to look for the greatest value or the least value is limited.

D. Fermat found simple computational formulas for determining the slope of the tangent line, today called the *derivative* of the function. These formulas included rules for the derivative of any polynomial and for the derivative of an arbitrary rational power of the variable.

V. Fermat was also interested in perfect numbers and Pythagorean triples.

A. A perfect number is one that is equal to the sum of its proper divisors. For example, the sum of the proper divisors of 6—1, 2, and 3—is 6. Perfect numbers were important to the Greeks and later thinkers. Four perfect numbers were known to the Greeks: 6, 28, 496, and 8128.

B. All even perfect numbers are constructed by first finding a prime that is 1 less than a power of 2. For example, 6 is 2×3, and 3 is 1 less than a power of 2.

C. Fermat did important work in understanding when a number that is 1 less than a power of 2 can be a prime.

1. First, the exponent must be a prime: $2^8 - 1$ cannot be prime; $2^9 - 1$ cannot be prime; $2^{11} - 1$ might be prime. In fact, it's not ($2^{11} - 1 = 23 \times 89$), but it raises an interesting point.

2. The exponent in the last example is 11. Both 23 and 89 are 1 more than a multiple of 11. Fermat realized that if we raise 2 to a prime power and subtract 1, the only numbers that can divide into the result are 1 more than a multiple of the prime exponent.
3. This discovery would come to be known as *Fermat's little theorem*. It is the basis of the modern RSA public key cryptosystem that is used for secure communication over the Internet.
4. The problem of finding prime numbers equal to 1 less than a power of 2 would continue to interest mathematicians. These numbers are called *Mersenne primes* in honor of Father Mersenne; the largest prime known today is of this form: $2^{32,582,657} - 1$.

D. In looking at Pythagorean triples (integers such as 3, 4, and 5, where $3^2 + 4^2 = 5^2$), Fermat wondered if he could find three positive integers that met the same condition when raised to a higher power (for example, $a^3 + b^3 = c^3$).
1. In the margins of his copy of *Arithmetica* by Diophantus, Fermat wrote: "It is impossible … in general, for any number which is a power greater than the second to be written as a sum of two like powers. I have a truly marvelous demonstration which this margin is too narrow to contain."
2. Fermat almost certainly did not have a proof of this result, which came to be known as *Fermat's last theorem*. This assertion, in fact, would not be settled until the end of the 20th century.

Suggested Readings:

Edwards, *Fermat's Last Theorem*, chap. 1.

Gindikin, *Tales of Mathematicians and Physicists*, 129–50.

Katz, *A History of Mathematics*, 432–42, 448–60, 470–73, 481–85.

Van der Waerden, *A History of Algebra*, chap. 3.

Questions to Consider:

1. Certain ideas seem to wait until the time is ripe for discovery, and then several people hit on them simultaneously. This is the case with analytic geometry. What were the pieces that needed to

be in place before this means of translating between algebra and geometry could be discovered?

2. Richard Hamming has commented that he finds it intriguing that the simplest ideas in algebra—linear expressions, then quadratic expressions—correspond to the simplest ideas in geometry—straight lines, then conic sections. Is this surprising? Should it be?

Lecture Twelve—Transcript

Fermat, Descartes, and Analytic Geometry

In this lecture we're going to look at Pierre de Fermat and also René Descartes. Fermat really is instrumental in the development of analytic geometry and the origins of calculus. This will be the first time that I talk a little bit about Fermat's last theorem. Fermat was by profession a lawyer whose mathematics was simply an avocation. Fermat had been born in 1601; he would die in 1665. Born near Toulouse, he first went to Bordeaux, and there he actually studied mathematics—studied with some of the disciples of Viète, a French algebraist from the 1500s who was no longer alive at that time, but people who had studied with Viète were around. Fermat learned a lot of his mathematics at that time.

From Bordeaux, he then went on to Orléans, where he got a law degree and then returned to Toulouse. He initially got a job as counselor to Parliament, and it was in this capacity as a counselor that he was able to add the "de" to his name, and so he became Pierre de Fermat. He fairly quickly rose through the ranks as a lawyer in service to the Parliament and to the government in Toulouse. By 1638, he was counselor for the upper house of Parliament, and by 1652, he was working for the criminal court and would eventually attain one of the highest positions there. In fact, his rise through the law profession was so meteoric that a number of people have postulated that it may have been his good luck that the plague came through Toulouse so frequently—that his superiors probably were dying off, which is why he was able to rise through the ranks. He himself at one time did actually contract the plague, and the rumor went out for a while that he had actually died of the plague, though fortunately for us he did survive.

As I said, Fermat looked at mathematics as an avocation. He actually published very little in his lifetime, and we might not even know about Pierre de Fermat if it had not been for his son. It was his son who would collect his manuscripts after his father's death and then see that they got published and that his name was more widely known. The first mathematics that Pierre de Fermat did came out of work that he had begun in the 1620s while he was in Bordeaux. He had come across a description of a lost book from Apollonius. This is Apollonius of Perga, the author of the *Conics*. One of the books that he wrote was called *Plane Loci*, which dealt with two-dimensional curves. That book is lost to us, but a later Greek mathematician by the name of Pappas had written about this particular book of Apollonius and described many of the results that Apollonius had proven in his book.

Fermat took a translation into Latin of this description of the lost book of Apollonius, and he sought to try to prove these results on geometry. They were extremely difficult to prove, but in order to find a proof, what Fermat did was to invent an entirely new way of approaching geometric problems—what today we call *analytic geometry*. What he did was to take a geometric figure and reinterpret it algebraically. Usually the way we think of that today is to take an algebraic expression and look at it geometrically. When we graph an algebraic expression, we put a horizontal axis and we put a vertical axis. We've got our two variables, the x and the y. For each pair of values—a value for x and a value for y—that correspond to a point that satisfies an algebraic equation, we take a look at the x value and the y value, and we plot that point.

As you plot all of the points on what today is called the *Cartesian coordinate axis*, you get a curve traced out. You can actually look at this both ways. You can start with an algebraic expression and turn it into a geometric curve. You can also start with a geometric curve and turn it into an algebraic expression. This is exactly what Fermat did. He took the geometric statements that Apollonius had made, and he translated them into algebraic statements; then, he used this very advanced work on algebra that had been done in the preceding centuries in order to find simple solutions and simple proofs to these geometric problems.

Now when Fermat did his coordinate axes, he didn't do them quite the way we would do them today. He did not actually have a vertical axis. He had the horizontal axis on which one of the variables would be represented, and then he simply marked off the distance to the other variable—very much along the lines of what Galileo did, but remember that Galileo actually drew little vertical lines. One of Fermat's inventions was not to draw those little vertical lines, but simply to mark the point above the horizontal axis. He didn't actually put in the vertical axis, and actually as he found the distance from the horizontal axis, he didn't always move vertically. Sometimes he moved off at an angle. The important thing is that you're always moving off at exactly the same angle. It provides a method for translating this algebraic expression into geometric—or geometric back into algebraic. It's an idea that Fermat came up with. Descartes would also come up with the same idea at the same time.

Fermat had started on this work in the 1620s. It would be 1637 before he was ready to publish it. He published it in that year as the book *Introduction to Plane and Solid Loci*. Exactly that same year, by one of the great coincidences that often happens in mathematics, René Descartes independently came up with exactly the same idea for translating between geometric objects and algebraic expressions. René Descartes published his great work, *Geometry*, contained in a more comprehensive work called *Discourse on the Method for Rightly Directing One's Reason and Searching for Truth in the Sciences*. One of the things that you cannot accuse Descartes of is being overly humble. He believed that he had found the right way for answering any scientific question that would be out there.

In fact, Descartes' explanation of how to pass back and forth between geometric and algebraic expressions would become much more popular than Fermat's for a couple of reasons. One, Descartes was just much better known. Also, Descartes was very good at coming up with efficient algebraic notation. The notation that we use today is really Descartes' algebraic notation. Fermat was using an algebraic notation that he had learned from Viète's disciples. It's not as efficient; it's not as easy to work with. A lot of people looked at Fermat's work, and they couldn't understand what he was really doing, or they had trouble reading it. That is why today when we talk about a coordinate plane, we talk about it as a "Cartesian plane," named after Descartes, rather than a "Fermatian plane," named after Pierre de Fermat—although in reality, both of them discovered the same idea and published this idea in exactly the same year.

Now there is an important difference between how Descartes and Fermat looked at this problem of moving between geometry and algebra. Descartes, like Fermat, started with this idea of taking a geometric object, interpreting it algebraically, and then being able to come to geometric conclusions by studying the algebra. That is how Fermat began this, but Fermat fairly quickly realized that he could move in the other direction, and that would be very powerful. He could take an algebraic expression—say, the equation that describes a quadratic polynomial, a polynomial of degree two—and he could then graph that algebraic expression and work with a parabola instead of the algebraic expression. When you take an algebraic expression and you look at it geometrically, very often you can see things in the geometry that are not immediately apparent when you're looking at the algebra.

Two examples of this are—you can think about the area underneath the geometric object—so, I've got a parabola, perhaps a parabola that is opening downward—and one can ask: What is the area between the horizontal axis and this particular parabola? The other thing that you can look at is: Where does this algebraic expression hit its largest value? Where does it reach its smallest value, say within some range of values? If you look at it geometrically, the highest value is simply going to be where that curve peaks, and the lowest value is going to be where that curve has its lowest valley. That ability to take an algebraic expression and translate it into something geometric would very directly lead into the development of much of calculus. This is something, unfortunately, that Descartes missed.

I want to say just a little bit more about Descartes. I mentioned before that he is the person who invented the term *imaginary number* for the square root of a negative number—again, a very, very unfortunate term because these are numbers that truly exist even though Descartes called them "imaginary." This is also a time in the 17th century when people were still not sure about whether to accept negative numbers as legitimate numbers.

Now, I've talked about the introduction of negative numbers. The Chinese were the first to work with them. They didn't accept them as legitimate solutions, but they realized that they were important as intermediate steps. The Indian mathematicians eventually would adopt them not just as intermediate steps, but actually as legitimate answers. The Islamic mathematicians were very much divided on whether or not to allow negative numbers. They were certainly very useful. On the other hand, they were very much embedded in the Greek tradition and the Babylonian tradition that looked at numbers as describing something. Ideally, numbers were describing lengths in some way. You can't talk about a negative length, and so it never really seemed to make sense to talk about negative numbers.

When we get into western Europe in the 15th, 16th, and 17th centuries, there was a very real ongoing debate about whether or not to accept negative numbers as legitimate answers. Descartes was one of the last to really speak out against accepting negative numbers. He would talk about a negative root of a polynomial (a negative solution to a polynomial equation) as a "false root" and a positive solution as a "true root." Fortunately, that is a terminology that we have long since lost. Certainly, by the middle of the 17th century, and well into the late 17th century, people were now accepting negative numbers as legitimate numbers that really could be answers to problems that were out there.

One of the other things that Descartes did that really set science back quite a bit was his *Principles of Philosophy*, published in 1644. His idea was to put out principles on which you could base all of your knowledge of the universe. He was looking at this problem that Aristotle had laid out, foundations for understanding the world around us, and the Aristotelian foundations were crumbling. Descartes decided to replace those with modern foundations that would hold up, so this is where we get such phrases as "*Cogito, ergo sum*" (I think, therefore I am). His work consists of four major volumes: *The Principles of Human Knowledge*, *The Principles of Material Things*, *Of the Visible World*, and *The Earth*.

One of the things that Descartes stated quite forcefully in this particular collection was that there can be no action without an actor—no action at a distance. If something is moving, it's because something else is moving it. He explained the motion of the planets by postulating an ether that pervaded all of the universe, and it was actually the swirling ether that was moving the planets in their courses. He explained the fact that we don't sense motion because we are embedded in this ether—we are moving with the ether. Since we're moving with this ether, we don't sense this motion. I said that would be one of the great disasters of the mid-17th century, and that is because when Newton published his *Principia*, he was looking at gravity. Gravity is something that acts at a distance, and Newton was unable to explain how it was possible for gravity to act at a distance without any intermediary. A lot of people criticized Newton—in fact, refused to accept his great work on celestial mechanics—precisely because he could not come up with a Cartesian-type mechanism in order to explain how gravity was able to act.

I said that Fermat published very little. Even though he published very little, he was an active correspondent in mathematics. Most of his correspondence was undertaken with a Catholic priest in Paris by the name of Father Marin Mersenne, a very important figure during the 1600s. This is a time before scientific journals, and so if you wanted to find out what someone else was doing, you had to learn it by letter. There were a few key people—and Father Mersenne was one of the most important of these—who would correspond with the scientists of that time. He would hear what different people were doing, what they had discovered, and then he would send out letters to all of his acquaintances describing this. He was a great friend of Descartes. He was a good friend of Christiaan Huygens, who I will talk about in the next lecture, and he also corresponded regularly with Fermat.

Mersenne, incidentally, was also a great supporter of Galileo, despite the fact that he was a Catholic priest. In secret, he managed to smuggle out a copy of Galileo's *Dialogue*, this work that had been banned by the Inquisition. It was Mersenne who managed to get a copy of this to the Netherlands and to see that it got published. Mersenne is known to have had over 78 correspondents during his life—including not just Descartes and Huygens and Fermat, but also Gassendi, Roberval, Pell, Torricelli, Hobbes, and both Etienne and Blaise Pascal—all of the great scientists of his time.

In 1636, Fermat tackles the problem of finding the area underneath a curve that is described by an algebraic expression. This is, in essence, what would come to be understood as the integral calculus. Given an algebraic expression, consider its curve—the curve that represents that expression—and look at the area underneath it. It would be Fermat who would be the first person who would actually discover how to find the area underneath this curve for an arbitrary algebraic expression—coming up with what today we think of as the means of integrating polynomials. He used precisely the method of exhaustion that goes back to Eudoxus of Cnidus, and that had been used then by Euclid, and Archimedes, and Ibn al-Haytham, and many mathematicians since then. Fermat showed how to use this idea in order to find areas under curves. He was not the only person. One of the other correspondents of Mersenne (in the very same year of 1636), Gilles de Roberval, also found exactly the same formula for the area under a curve that corresponds to an arbitrary polynomial.

Fermat then went on and considered the problem: Given an algebraic expression, how do we find where it reaches its maximum (its greatest value) or its minimum (its least value)? In 1639, he would discover how to do this. The idea was to take the algebraic expression, represent it as a curve, and then look at the line that is tangent to the curve at each point. If we consider the line that touches the curve at just a single point, if I've got an expression that is increasing, that slope is going to be positive. If I've got an expression that is decreasing, that slope is going to be negative. If I've got an expression that is at its highest value, or at its very lowest value, then that tangent line is going to be horizontal; its slope is going to be zero.

Fermat did not stop there. He then looked at how to determine the slope of the tangent line at a single point. What he did then was to consider two points that are very close to the point in question on the curve that we're looking at. It's possible to actually construct the equation of the straight line that goes through those two points, and the slope of the line that goes through those two points is going to be the rise over the run—it's the change in the y value divided by the change in the x value. Fermat's brilliant insight was that you could take this change in the x value to be smaller, and smaller, and smaller. As the change in the x value gets smaller, the change in the y value gets smaller, and what happens is that this ratio of two numbers—both the numerator and denominator are getting closer to zero—is going to approach some number that is the slope of the line that is precisely tangent at that particular point.

When we want to find the maximum value or the minimum value, we do precisely this procedure, but we're looking for those points where the slope of the tangent line is exactly equal to zero. Now, the fact that the slope of the tangent line is zero does not guarantee that we have either a maximum value or a minimum value. The function might increase, level off at a kind of step, and then start increasing again. But in general, for most algebraic expressions, you're going to have very few places where the function actually has a horizontal tangent line—so that greatly limits the number of places where you have to look for the greatest value or the least value. In fact, it restricts it so that you can then simply look at the value of the function at the different candidates and pick that value of the function at which you get the largest value or the smallest value.

Fermat extended this idea to actually find ways of determining this slope of the tangent line, what today we call the *derivative* of the function. Fermat found rules for doing this for any polynomial, for arbitrary—not only integer—powers of an unknown, but also for arbitrary rational powers of an unknown. So, if I've got a curve that's representing $\sqrt{x}$, or $\sqrt[3]{x}$, or $x^{5/3}$, Fermat showed how to find the slope of the tangent line for an expression such as that.

I want to spend the rest of this lecture looking at Fermat's work in number theory, because that would be extremely important. Fermat read Diophantus's *Arithmetica*, and one of the things that he discovered there was the Pythagorean triples. Something else that he discovered in the work of Diophantus was perfect numbers. A perfect number is one that, if you take the sum of its proper divisors, gets you back to that number. So, in the case of 6, the sum of its proper divisors—1, 2, and 3—is 6 again. If I take the proper divisors of 28, they are 1, 2, 4, 7, and 14. If I add those numbers up, I get back to 28. Perfect numbers were important to the Greeks. They continued to be important. In fact, one of my favorite quotes comes from St. Augustine of Hippo. In *The City of God*, he muses on why it took God exactly six days to complete the creation, and he says in that: "It is recorded that all God's works were completed in six days, because six is a perfect number."

There were four perfect numbers that were known to the ancient Greeks: 6, 28, 496, and 8128. Fermat was interested in this problem of going beyond that. One of the keys to finding a perfect number, as Fermat realized, was to find a prime that is 1 less than a power of 2. So, if I look at 6, that's 2×3, and 3 is 1 less than a power of 2. If I look at 28, that is 4×7, and 7 is 1 less than a power of 2. The even perfect numbers—we actually don't know if there are any odd perfect numbers or not—all have to be of this form. It's going to be a power of 2 times a prime that is 1 less than a power of 2.

Fermat did very important work in understanding when a number that is 1 less than a power of 2 can be a prime. First of all, the exponent has to be a prime: $2^8 - 1$ cannot possibly be prime; $2^9 - 1$ cannot possibly be prime; $2^{11} - 1$ might be a prime. In fact, it's not ($2^{11} - 1 = 23 \times 89$), but something interesting is happening there. If I look at 23 and 89, I was using the exponent 11. Both 23 and 89 are 1 more than a multiple of 11. There is something very deep going on here that Fermat realized—that if I take a prime, take 2 to that prime power and subtract 1, the only numbers that can divide into that are numbers that are 1 more than a multiple of that prime exponent. This would come to be known as *Fermat's little theorem*, and it would come to play a very important role in number theory. In fact, the RSA public key cryptosystem that I'll talk about in the very last lecture has its foundation in this observation of Fermat.

This problem of when is 1 less than a power of 2 equal to a prime is one that would continue to interest mathematicians. In fact, the very largest known prime today is of this form. It is called a *Mersenne prime* in honor of Father Marin Mersenne, and the largest prime known at the moment that I am speaking—and this changes about once a year—is the 44th prime that is 1 less than a power of 2. It is $2^{32,582,657} - 1$. It's an enormous number. It's got almost 10 million digits in it, so I'm not going to be able to show you that number. But in general, the largest known prime has usually been one of these Mersenne primes, a number that is 1 less than a power of 2.

Fermat also was interested in the problem of Pythagorean triples (integers like 3, 4, 5, where $3^2 + 4^2 = 5^2$; or 5, 12, 13, where $5^2 + 12^2 = 13^2$). That led him to think, what about using higher powers? Can I find three positive integers so that the cube of the first plus the cube of the second is equal to the cube of the third ($a^3 + b^3 = c^3$)—or use fourth powers; the fourth power of the first plus the fourth power of the second is equal to the fourth power of the third ($a^4 + b^4 = c^4$)? He was reading Diophantus, his *Arithmetica*, at this time, and he was writing his notes in the margin. This copy of Diophantus with the marginal notes is something that Fermat's son would publish with all of these marginal notes. This is how we come to know of what today is called *Fermat's last theorem*. Fermat wrote in the margin, thinking about doing Pythagorean triples, but for higher powers:

> It is impossible for a cube to be written as a sum of two cubes or a fourth power to be written as a sum of two fourth powers or, in general, for any number which is a power greater than the second to be written as a sum of two like powers. I have a truly marvelous demonstration which this margin is too narrow to contain.

Almost certainly, he did not have a proof of this result, which came to be known as Fermat's last theorem—called "Fermat's last theorem" because over the years people would prove all of the other results that Fermat had stated without proof, or show that they were wrong. This was the last of his assertions that needed to be settled, and it would not be finally settled until the very end of the 20th century, but that is a topic for a later lecture. For the next lecture, we're going to turn to Isaac Newton and the full development of calculus.

Timeline

B.C.

2030–1640 .... Egyptian Middle Kingdom.

2000–1600 .... Old Babylonian period.

800–200 .... Composition of the Indian *Sulbasutras*.

c. 624–c. 545 .... Life of Thales of Miletus.

c. 563–483 .... Life of Siddhartha Gautama, the Buddha.

551–479 .... Life of Confucius.

c. 520 .... Pythagoras founds his school in Samos.

388 .... Plato founds the Academy in Athens.

323 .... Alexander the Great dies; Ptolemy I rules Egypt; Seleucus I rules Persia and Mesopotamia.

c. 300 .... Museion at Alexandria is founded, Euclid flourishes.

3rd century .... Period of the Warring States in China.

212 .... Rome conquers Syracuse; death of Archimedes.

208 .... Founding of Han dynasty in China.

c. 200 .... Apollonius of Perga writes the *Conics*.

127–126 .... Hipparchus makes his astronomical observations in Rhodes.

30 .... Rome annexes Hellenistic Egypt.

A.D.

1st century .... Buddhism enters China; Kushan Empire extends into northern India.

early 2nd century .... Ptolemy writes the *Almagest*.

mid-3rd century .... Diophantus writes *Arithmetica*.

late 3rd century .... Liu Hui writes *Sea Island Computational Canon*.

c. 320 .... Founding of Gupta Empire in India.

337 .... Roman Emperor Constantine I baptized as a Christian.

391 .... Theodosius I orders the destruction of all pagan temples in the Roman Empire; probable end of the Museion in Alexandria.

415 .... Death of Hypatia of Alexandria.

late 5th century .... Zu Chongzhi discovers $\frac{355}{113}$ as approximation to π.

618 .... Founding of Tang dynasty in China.

622 .... Muhammad flees from Mecca to Medina: beginning of Islamic calendar.

644–648 .... Li Chunfeng collects and revises existing Chinese mathematical treatises into the *Ten Computational Canons*.

mid-7th century .... Brahmagupta leads the astronomical observatory at Ujjain.

750 .... Abbasid caliphate founded in Baghdad.

786–809 .... Harun al-Rashid rules in Baghdad.

825 .... Al-Kwarizmi writes his treatise on *al-jabr* and *al-muqabala*.

c. 1000 .... Al-Haytham begins his work as scientist and engineer in Cairo.

mid-11th century .... Jia Xian produces first definitively known example of "Pascal's triangle."

1088 .... Founding of the University of Bologna.

1149 .... Al-Samawal writes *The Brilliant in Algebra*.

12th century .... Bhaskara Acharya leads the observatory at Ujjain.

1202 .... Leonardo of Pisa (Fibonacci) writes the *Liber abaci*.

1235 .... Ujjain conquered by the Delhi caliphate and destroyed.

1258 .... Hulagu Khan sacks Baghdad.

1260–1294 .... Rule of Kublai Khan in China.

1303 .... Zhu Shijie writes *Trustworthy Mirror of the Four Unknowns*.

c. 1450 .... Gutenberg invents moveable-type printing.

1453 .... Constantinople conquered by Ottoman Turks.

1492 .... Ferdinand and Isabella conquer Granada; Columbus lands in America.

c. 1505 .... Del Ferro discovers method for finding roots of arbitrary cubic polynomial.

1517 .... Luther nails his theses to church door in Wittenberg.

1525 .... Albrecht Dürer publishes his book on geometric constructions.

1543 ....Nicolaus Copernicus publishes *De revolutionibus*.

1545 ....Gerolamo Cardano publishes *Ars magna*.

1581 ....Dutch Republic wins freedom from Spain.

1585 ....Simon Stevin publishes *La Thiende*, advocating the use of decimal fractions.

1601 ....Tycho Brahe dies in Prague; his assistant, Johannes Kepler, inherits his astronomical data.

1607 ....Jamestown is founded in Virginia.

1610 ....Galileo publishes *The Starry Messenger*.

1614 ....Napier publishes *Description of the Wonderful Canon of Logarithms*.

1620 ....Plymouth Colony is established in Massachusetts.

1637 ....Fermat and Descartes publish descriptions of analytic geometry.

1642 ....Death of Galileo; birth of Newton.

1644 ....Descartes publishes *Principles of Philosophy*.

1653–1658....................................Cromwell rules as Lord Protector.

1655 ....Wallis publishes *Arithmetic of Infinities*.

1660 ....Royal Society of London is founded.

1663 ....Barrow becomes first to hold the Lucasian Chair in Mathematics at Cambridge.

1684 ....Leibniz publishes his first paper on calculus.

1687 ....Newton publishes *Mathematical Principles of Natural Philosophy*.

1700 ....Prussian Academy of Sciences is founded.

1713 ....Jacob Bernoulli's posthumous *The Art of Conjecture* is published by his brother, Johann.

1727 ....Euler joins the newly founded St. Petersburg Academy of Sciences.

1744 ....Frederick the Great reestablishes the Royal Academy of Sciences in Berlin.

1748 ....Euler publishes *Introduction to Analysis of the Infinite*.

1789 ....Storming of Bastille, start of French Revolution.

1794 ....Founding of the École Normale.

1799–1825....................................Laplace publishes *Treatise on Celestial Mechanics*.

1801 ....Gauss publishes *Investigations of Arithmetic*.

1807 ....Fourier submits his thesis on *The Propagation of Heat in Solid Bodies*.

1815 ....Defeat of Napolean at Waterloo; reinstitution of the Bourbon monarchy.

1821 ....Cauchy publishes *Cours d'analyse*.

1824 ....Abel proves impossibility of solving the general quintic.

1830 .... Charles X flees France; accession of Louis-Philippe.

1832 .... Galois is killed in duel.

1848 .... End of French monarchy, beginning of rule of Napoleon III.

1848–1870.................................. Independence and unification of Italy.

1863–1871.................................. Unification of Germany.

1864 .... Riemann delivers his lecture "On the hypotheses that lie at the foundations of geometry."

1865 .... Maxwell publishes *A Dynamical Theory of the Electromagnetic Field*.

1878 .... Sylvester founds *American Journal of Mathematics*; Cayley publishes *The Theory of Groups*.

1887 .... Hertz detects electromagnetic potential.

1889 .... Kovalevskaya becomes first woman to hold a professorship in mathematics at a European university.

1892–1899.................................. Poincaré publishes *Lectures on Celestial Mechanics*.

1896 .... Hadamard and de la Vallée Poussin independently prove the prime number theorem.

1900 .... Hilbert announces his 23 problems.

1914 .... Ramanujan travels to England to work with Hardy.

1914–1918.................................. World War I.

1915–1916 .... Einstein publishes the theory of general relativity.

1922 .... De Broglie introduces the wave-particle duality of electrons.

1936 .... Ahlfors and Douglas are first recipients of the Fields Medal.

1939–1945 .... World War II.

1946 .... ENIAC computer.

1957 .... Sputnik.

c. 1970 .... Early development of string theory.

1984 .... First Macintosh computer.

1991 .... Debut of the World Wide Web.

1995 .... Wiles publishes proof of Fermat's last theorem.

2000 .... Clay Mathematics Institute announces the seven Millenium Prize Problems.

2002 .... Perelman announces proof of the Poincaré conjecture.

2003 .... Serre becomes first recipient of Abel Prize; completion of the Human Genome Project.

Glossary

algebra: The field of mathematics that deals with expressions in unspecified quantities. In its purest sense, it seeks to find the value of an unknown quantity by manipulating a balance of two different expressions in that unknown quantity. In the 19th century, the meaning of the term expanded to cover generalized number systems and the study of transformations and symmetries.

amicable numbers: A pair of numbers, such as 220 and 284, for which the sum of the proper divisors of one equals the other.

analysis: As a field of mathematics, the advanced study of calculus that appeared in the 19th century.

analytic geometry: The area of mathematics that combines algebra and geometry by representing algebraic expressions as geometric curves plotted on a Cartesian plane.

calculus: A field of mathematics with its origins in the problems of calculating slopes of tangent lines and rates of change (differential calculus) as well as areas, volumes, and other quantities that are limits of sums of products (integral calculus).

chord: The straight-line segment connecting any two points on a circle.

combinatorics: The mathematics of counting arguments.

complex number: Any sum of a real and an imaginary number. Complex numbers represent points in a plane.

continued fraction: A representation of a number as a sequence of integers obtained by specifying the integer part of the number and then iteratively finding the integer part of the reciprocal of what remains.

decimal system: A place-value system based on powers of 10.

degree: $\frac{1}{360}$ of the circumference of a circle.

digit: Any of the 10 symbols 0 through 9.

Diophantine equation: Any equation for which possible solutions are restricted to integers.

elliptic function: A doubly-periodic complex-valued function of a complex variable.

elliptic integral: Any of a family of integrals among which is the integral for determining the length of the arc of an ellipse.

Fermat's last theorem: The statement made by Fermat that for any exponent $n > 2$, there is no triple of positive integers *x*, *y*, *z* for which $x^n + y^n = z^n$.

Fourier series: The representation of a function by an infinite series of trigonometric functions.

fundamental theorem of algebra: The statement that every polynomial with real coefficients has at least one root (which might be complex).

fundamental theorem of calculus: The statement of the equivalence of two ways of thinking of integration: as the inverse process of differentiation and as a limit of sums of products.

geometry: The mathematical abstraction of spatial relationships.

harmonic series: The sum of the reciprocals of the positive integers.

imaginary number: The square root of a negative number.

infinite series: An infinite summation.

irrational number: A number that cannot be represented as a ratio of two integers.

isochrone: A curve with the property that a ball placed anywhere along it will take the same amount of time to reach the bottom.

logarithm: An exponent. Specifically, given a number larger than one, called the *base*, the logarithm of any positive number is the power of the base that will yield that number.

method of exhaustion: A means of finding areas and volumes by breaking the region up into ever-finer pieces.

minute (′): $\frac{1}{60}$ of a degree.

modular function: A function that exhibits invariance under a set of transformations of the independent variable that includes translations and the taking of the negative of the reciprocal.

number theory: The study of the integers, especially the pursuit of integer solutions to mathematical problems.

Pascal's triangle: A triangular arrangement of the coefficients of the expansions of powers of the binomial $1+x$.

perfect number: A number, such as 6 or 28, that is equal to the sum of its proper divisors.

pi (π): The ratio of the circumference of a circle to its diameter.

place-value system: A method for recording numbers that assigns different values to the digits depending on their position. Thus, the 2 in 27 represents two 10s, or 20.

Platonic solids: The five solids whose faces are identical regular polygons.

polynomial: A mathematical expression involving a sum of powers of a variable in which each term may be multiplied by a number, such as x^2-3x+6.

Pythagorean theorem: The statement that in any right triangle, the sum of the areas of the squares whose sides are adjacent to the right angle is equal to the area of the square whose side is opposite the right angle.

Pythagorean triple: Three positive integers that form the sides of a right triangle.

rational number: A number that can be represented as a ratio of two integers.

Riemann hypothesis: The statement that the nonreal zeroes of the zeta function, a function used to study the distribution of prime numbers, all have real parts equal to $\frac{1}{2}$.

root: A root of a polynomial is a value at which the polynomial is zero.

second (″): $\frac{1}{60}$ of a minute.

sexagesimal system: A place-value system based on powers of 60.

sine: A half-chord.

square root: Given a number that represents an area, the square root of that number is the length of the side of the square with that area.

statics: The physics of counterbalancing forces.

Sulbasutras: Appendices to the Vedas that include detailed mathematical descriptions of altar construction.

topology: A mathematical field within geometry that studies properties that are left unchanged by small changes in how distance is measured. Informally known as *rubber sheet geometry*.

trigonometry: The mathematics built on the study of chords.

variable: An unspecified quantity that can be assigned more than one value.

Vedas: Hindu mythological texts composed over the period 2500–600 B.C.

Biographical Notes

Abel, Niels Henrik (1802–1829). Norwegian mathematician who was first to prove the impossibility of a general solution to polynomial equations of degree five.

Al-Haytham (c. 965–1040). Mathematician, scientist, and engineer born in Basra; worked in Baghdad and Cairo. Noted for his work in optics and his methods for finding volumes of solids of revolution.

Al-Kwarizmi (c. 790–840). Baghdad mathematician whose *Condensed Book on the Calculation of Restoring and Comparing* is considered the first book of algebra.

Al-Samawal (c. 1130–c. 1180). Jewish doctor and scientist from Baghdad who made important advances in algebra.

Apollonius of Perga (c. 260–c. 190 B.C.). The inventor of the astronomical system of epicycles and the author of the *Conics*, which established the mathematical properties of parabolas, ellipses, and hyperbolas.

Archimedes of Syracuse (287–212 B.C.). The greatest of the Greek scientists. Among his many accomplishments was the development of the method for finding areas and volumes.

Aryabhata (476–550). Indian astronomer who worked at Kusumapura, near modern Patna.

Bernoulli, Jakob (1654–1705). Swiss scientist and author of *The Art of Conjecture*, the founding work in the theory of probability.

Bernoulli, Johann (1667–1748). Swiss scientist who worked with Leibniz, helped establish calculus, and taught Leonhard Euler.

Bhaskara Acharya (1114–1185). Last of the great astronomers at Ujjain, India. Discovered general methods for finding quadratic approximations and for solving the problem today known as Pell's equation.

Bombelli, Rafael (1526–1572). Important algebraist who introduced and explained the workings of complex numbers.

Brahmagupta (598–c. 665). Head of the astronomical observatory at Ujjain, India. Made significant contributions to trigonometry and number theory.

Cantor, Georg (1845–1918). Founder of modern set theory.

Cardano, Gerolamo (1501–1576). Preeminent algebraist of the 16th century and author of *Ars magna* (*The Great Art*).

Cauchy, Augustin-Louis (1789–1857). French mathematician; considered the founder of analysis.

Cayley, Arthur (1821–1895). British mathematician who began the unification of non-Euclidean and projective geometries and helped to establish modern algebra.

Cohen, Paul (1934–2007). The American mathematician who finally established the status of the axiom of choice and the continuum hypothesis.

Diophantus of Alexandria (c. 200–284). The first to introduce algebraic notation; author of *Arithmetica*, the book that would inspire much future work in number theory.

Euclid of Alexandria (c. 325–c. 265 B.C.). Hellenistic scholar who established the mathematical community at the Museion of Alexandria and who consolidated all Greek knowledge of mathematics.

Eudoxus of Cnidus (c. 395/390–c. 342/337 B.C.). Greek mathematician who is credited with the discovery of the method of exhaustion for finding areas.

Euler, Leonhard (1707–1783). Swiss mathematician who worked in St. Petersburg and Berlin. One of the most prolific mathematicians and probably the most influential.

Fermat, Pierre (1601–1665). Councilor to the parliament in Toulouse, inventor of analytic geometry, and one of the important contributors to the development of calculus and number theory.

Fontana, Niccolò (aka **Tartaglia**; 1499–1557). Independent discoverer of the technique for finding the exact value of a root of any cubic equation.

Fourier, Joseph (1768–1830). French mathematician, scientist, and bureaucrat.

Galilei, Galileo (1564–1642). Italian mathematician and scientist noted for his attempts to explain the physics of motion.

Galois, Evariste (1811–1832). French mathematician; first person to solve the general problem of when a polynomial equation can be solved exactly.

Gauss, Carl Friedrich (1777–1855). German astronomer and mathematician who spent most of his career at Göttingen and whose contributions span geometry, number theory, and analysis.

Germain, Sophie (1776–1831). French mathematician who made important contributions to the understanding of Fermat's last theorem.

Gödel, Kurt (1906–1978). Austrian logician who proved the incompleteness theorem.

Hilbert, David (1862–1943). German mathematician who made important contributions in algebra, geometry, and analysis.

Hipparchus of Rhodes (190–120 B.C.). Astronomer and father of trigonometry.

Huygens, Christiaan (1629–1695). Dutch scientist who invented the pendulum clock, discovered a Moon of Saturn, and served as mentor to both Leibniz and Johann Bernoulli.

Hypatia of Alexandria (c. 370–415). The first woman known to have made important contributions to mathematics, she wrote commentaries on several of the classic texts.

Jia Xian (fl. mid-11th century). Chinese mathematician; first person known to have recorded Pascal's triangle.

Klein, Felix (1849–1925). Head of the mathematics faculty at Göttingen and leader in the development of geometry.

Kovalevskaya, Sofya (aka **Sonya**; 1850–1891). Russian mathematician who worked in analysis and taught at the University of Stockholm.

Lebesgue, Henri (1875–1941). French mathematician known for his radically different approach to integration.

Leibniz, Gottfried Wilhelm (1646–1716). Librarian to the Duke of Hanover; shares with Newton the claim to be one of the founders of calculus.

Leonardo of Pisa (aka **Fibonacci**; c. 1170–1240). Merchant from Pisa who learned Islamic mathematics and wrote several important books in which he shared his knowledge.

Liu Hui (fl. late 3rd century A.D.). Earliest known Chinese mathematician; author of the *Sea Island Computational Canon*.

Napier, John (1550–1617). Scottish nobleman and theologian; inventor of the logarithm.

Newton, Isaac (1642–1727). Lucasian Professor of Mathematics at Cambridge and author of *Mathematical Principles of Natural Philosophy*.

Noether, Emmy (1882–1935). German algebraist.

Poincaré, Henri (1854–1912). French mathematician noted for his work in analysis and number theory and his work on the stability of complex systems.

Ptolemy of Alexandria (c. 100–c. 170). Astronomer and author of *Mathematiki Syntaxis*, which would come to be known as *Almagest*, the book that would shape astronomy until the 16th century.

Pythagoras of Samos (c. 580–c. 500 B.C.). Greek mystic who established a philosophical school at Croton, in what is now Italy, based on the premise that "all is number."

Ramanujan, Srinivasa (1887–1920). Self-taught Indian mathematician who worked on elliptic and modular functions.

Riemann, Bernhard (1826–1866). German mathematician who made seminal contributions to geometry, analysis, and number theory.

Stevin, Simon (1548–1620). Belgian scientist known for his contributions to algebra and statics and largely responsible for the European adoption of decimal fractions.

Sylvester, James Joseph (1814–1897). British algebraist; first professor of mathematics at The Johns Hopkins University and a founder of the American mathematical community.

Wallis, John (1616–1703). Savilian Professor of Geometry at Oxford and discoverer of many of the fundamental insights of calculus.

Weierstrass, Carl (1815–1897). Leading analyst of the mid-19th century; taught at the University of Berlin.

Wiles, Andrew (b. 1953). Princeton number theorist who proved Fermat's last theorem in 1995.

Zhu Shijie (c. 1260–1320). Chinese mathematician; author of *Siyuan yujian* (*Trustworthy Mirror of the Four Unknowns*).

Zu Chongzhi (fl. late 5th century A.D.). Chinese mathematician who discovered $\frac{355}{113}$ as an extremely accurate approximation to π.

Bibliography

Apollonius of Perga. *Conics*. Books I–III. Translated by R. Catesby Taliaferro. Reprint with corrections. Santa Fe, NM: Green Lion Press, 2000 (1939). A translation of the classic text on conic sections (circles, ellipses, parabolas, and hyperbolas) from the 3rd century B.C.

Bashmakova, Isabella, and Galina Smirnova. *Dolciani Mathematical Expositions*. Vol. 23, *The Beginnings and Evolution of Algebra.* Translated by Abe Shenitzer. Washington, DC: Mathematical Association of America, 2000. A comprehensive history from the geometric algebra of Babylon and Greece through the developments of the 19th century.

Bressoud, David. "Was Calculus Invented in India?" *College Mathematics Journal* 33 (2002): 2–13. A description of the Indian development of trigonometry and how questions of interpolation led to the invention of the derivative and infinite series.

Buck, R. Creighton. "Sherlock Holmes in Babylon." *The American Mathematical Monthly* 33 (1980): 335–45. An explanation of the decipherment of the Pythagorean triples in Plimpton 322.

Carlson, J., et al., eds. *The Millennium Prize Problems*. Cambridge, MA: Clay Mathematics Institute, 2006. Essays describing the seven millennium prize problems and summarizing the current state of our knowledge.

Cartwright, Mary L. "Mathematics and Thinking Mathematically." *The American Mathematical Monthly* 77 (1970): 20–28. Reprinted in *Musings of the Masters: An Anthology of Mathematical Reflections*, edited by Raymond G. Ayoub, 3–16. Washington, DC: Mathematical Association of America, 2004. Reflections on the sources of mathematics by one of the preeminent mathematicians of the 20th century.

Cohen, I. Bernard. *The Birth of a New Physics*. Rev. ed. New York: W. W. Norton & Company, 1985. An accessible explanation of the events leading up to Newton's *Principia* in terms of the development of our understanding of the physical world.

Courant, Richard, Herbert Robbins, and Ian Stewart. *What Is Mathematics?* Rev. ed. New York: Oxford University Press, 1996

(1941). A gentle and engaging introduction to the diversity of exploration of pattern that is mathematics.

Datta, B., and A. N. Singh. "Hindu Trigonometry." *Indian Journal of History of Science* 18 (1983): 39–108. An account of the trigonometric results that were known to Indian astronomers from A.D. 300 to 1500.

Dijksterhuis, E. J. *Archimedes*. Translated by C. Dikshoorn. Bibliographic essay by Wilbur R. Knorr. Princeton, NJ: Princeton University Press, 1987 (1938). The authoritative if now slightly dated account of Archimedes's mathematics.

———. *The Mechanization of the World Picture: Pythagoras to Newton*. Translated by C. Dikshoorn. Princeton, NJ: Princeton University Press, 1986 (1950). Originally published in Dutch, this source is now slightly outdated but is still the most thorough and erudite description of the development of our understanding of the physical world up to Newton's *Principia*.

Dunham, William. *Euler: The Master of Us All*. Washington, DC: Mathematical Association of America, 1999. Accessible explanations of much of Euler's mathematics.

Dyson, Freeman. "Missed Opportunities." *Bulletin of the American Mathematical Society* 78 (1972): 635–52. This is Dyson's Josiah Willard Gibbs Lecture to the annual meeting of the American Mathematical Society, in which he explains the essential nature of the connection between physics and mathematics: the idea that mathematics is essential to the building of physical understanding but physics is also essential to the development of new concepts in mathematics.

Edwards, C. H., Jr. *The Historical Development of the Calculus*. New York: Springer-Verlag, 1979. A collection of historically important discoveries related to calculus, starting with Archimedes and concluding with a brief overview of some topics from the 20th century. This book is written at a level to be used to supplement an undergraduate course in calculus.

Edwards, Harold M. *Fermat's Last Theorem: A Genetic Introduction to Algebraic Number Theory*. New York: Springer-Verlag, 1977. A graduate-level textbook explaining some of the mathematics to come out of the search for a proof of Fermat's last theorem. Although most of the book is written for an audience with a working knowledge of modern algebra, the first chapter describing what Fermat actually did is very accessible.

Euclid. *Elements*. Translated by Thomas Heath. New York: Dover Publications, 1956. The full 13 books of Euclid's *Elements*, still worthy of study as a foundation for mathematics.

Gillings, Richard J. *Mathematics in the Time of the Pharaohs*. Reprint, New York: Dover Publications, 1982 (1972). A general introduction to Egyptian mathematics.

Gindikin, Simon. *Tales of Mathematicians and Physicists*. Translated by Alan Shuchat. New York: Springer-Verlag, 2007. Fifteen essays on mathematicians and physicists from Gerolamo Cardano to Roger Penrose.

Gleick, James. *Isaac Newton*. New York: Pantheon Books, 2003. Newton's life and times; an insightful biography of the man and his accomplishments.

Gray, Jeremy. *Worlds Out of Nothing: A Course in the History of Geometry in the 19th Century*. London: Springer-Verlag, 2007. An undergraduate textbook and an excellent overview of this rich thread in mathematics.

Hamming, R. W. "The Unreasonable Effectiveness of Mathematics." *The American Mathematical Monthly* 87 (1980): 81–90. A reply to Wigner's article by one of the leading applied mathematicians of the 20th century.

Heath, Thomas. *A History of Greek Mathematics*. 2 vols. Reprint, New York: Dover, 1981 (1921). A thorough but accessible description of the history of Greek mathematics from Thales to Diophantus.

Høyrup, Jens. *Lengths, Widths, Surfaces: A Portrait of Old Babylonian Algebra and Its Kin*. New York: Springer-Verlag, 2002. A scholarly analysis of the geometric approach to algebraic questions used by the Babylonians.

James, Ioan. *Remarkable Mathematicians: From Euler to von Neumann*. Cambridge: Cambridge University Press, 2002. Brief biographies of 60 of the greatest mathematicians of the 18th, 19th, and first half of the 20th centuries.

Kanigel, Robert. *The Man Who Knew Infinity*. New York: Charles Scribner's Sons, 1991. A superb and accessible biography of Srinivasa Ramanujan.

Katz, Victor J. *A History of Mathematics: An Introduction*. 2nd ed. Reading, MA: Addison Wesley Longman, 1998. One of the best general histories of mathematics.

———. "Ideas of Calculus in Islam and India." *Mathematics Magazine* 68 (1995): 163–74. The discovery of formulas for sums of powers and their role in the development of integral calculus.

Klein, Jacob. *Greek Mathematical Thought and the Origin of Algebra*. Translated by Eva Brann. Reprint, New York: Dover Publications, 1992 (1969). Somewhat technical but offers an interesting description of Greek understandings of mathematical symbolism and how these were transformed by the European mathematicians of the 16th and 17th centuries.

Kline, Morris. *Mathematical Thought from Ancient to Modern Times*. New York: Oxford University Press, 1972. Now somewhat dated and occasionally flawed, but a great sweeping account of the history of mathematics that still provides inspiring insights for those curious about the nature of the discipline.

Laugwitz, Detlef. *Bernhard Riemann, 1826–1866: Turning Points in the Conception of Mathematics*. Translated by Abe Shenitzer. Boston, MA: Birkhäuser Verlag, 1999. A wonderfully thorough account of Riemann and his mathematics. Some of it gets technical, but most of the book can be read with little knowledge of advanced mathematics.

Markushevich, A. I. "Analytic Function Theory." In *Mathematics of the 19th Century*, edited by A. N. Kolmogorov and A. P. Yushkevich, translated by Roger Cooke, 119–272. Basel: Birkhäuser Verlag, 1996. A somewhat technical but primarily historical overview of the development of elliptic and modular functions and related questions in analysis.

Martzloff, Jean-Claude. *A History of Chinese Mathematics*. Translated by Stephen S. Wilson. Berlin: Springer-Verlag, 1997. A thorough and scholarly treatment of the history of Chinese mathematics before 1600, with a few references to Chinese mathematics into the 19th century.

Minio, Roger. "An Interview with Michael Atiyah." *The Mathematical Intelligencer* 6 (1984): 9–19. One of the great mathematicians of the 20th century talks about what mathematics means to him.

Naess, Atle. *Galileo Galilei: When the World Stood Still*. Berlin: Springer-Verlag, 2005. A modern and engaging telling of the story of Galileo and his struggle with the church.

Nasar, Sylvia, and David Gruber. "Manifold Destiny: A Legendary Problem and the Battle over Who Solved It." *The New Yorker*, August 28, 2007. An engaging account of the solution of the Poincaré conjecture and the politics of assigning credit for its solution.

Needham, Joseph. *Science and Civilisation in China*. Vol. 3, *Mathematics and the Sciences of the Heavens and the Earth*. Cambridge: Cambridge University Press, 1959. Now somewhat dated but still a useful overview of the history of Chinese mathematics to about 1600.

Nordgaard, Martin A. "Sidelights on the Cardano-Tartaglia Controversy." *National Mathematics Magazine* 13 (1937–1938): 327–46. The full dramatic story of the controversy over Cardano's right to publish the general solution of the cubic equation.

O'Shea, Donal. *The Poincaré Conjecture: In Search of the Shape of the Universe*. New York: Walker & Company, 2007. An engaging account of the Poincaré conjecture and its solution, written by a mathematician who truly knows this field.

Penrose, Roger. *The Road to Reality: A Complete Guide to the Laws of the Universe*. New York: Alfred A. Knopf, 2005. A massive and often overwhelming survey of the connection between mathematics and physics that proceeds at a dizzying pace—but it begins with the assumption that the reader has no more than a solid high school background in mathematics. It is one of the best introductions to the variety of models that attempt to unify the theories of general relativity and quantum mechanics.

Schattschneider, Doris. *M. C. Escher: Visions of Symmetry*. 2nd ed. New York: Harry N. Abrams, 2004. A delightful introduction to the mathematics at the heart of Escher's art.

Stein, Sherman. *Archimedes: What Did He Do Besides Cry Eureka?* Washington, DC: Mathematical Association of America, 1999. A popular account of Archimedes's principal mathematical results.

Straffin, Philip D., Jr. "Liu Hui and the First Golden Age of Chinese Mathematics." *Mathematics Magazine* 71 (1998): 163–81. An accessible introduction to Liu Hui's mathematics of the 3rd century A.D.

Swetz, Frank. "The Evolution of Mathematics in Ancient China," *Mathematics Magazine* 52 (1979): 10–19. A description for a general audience of several of the calculational techniques developed by Chinese mathematicians.

Toeplitz, Otto. *The Calculus: A Genetic Approach*. Translated by Luise Lange. Reprint, Chicago: University of Chicago Press, 2007 (1949). Still one of best introductions to the historical underpinnings of the ideas of calculus.

Van der Poorten, Alf. *Notes on Fermat's Last Theorem.* New York: John Wiley & Sons, 1996. An engaging introduction to many of the mathematical ideas connected to Fermat's last theorem, up to and including Wiles's proof. Written by a solid mathematician for an audience that is mathematically literate but not necessarily sophisticated. Those with a working knowledge of calculus have sufficient background.

Van der Waerden, B. L. *A History of Algebra from al-Khwarizmi to Emmy Noether*. Berlin: Springer-Verlag, 1985. The history of algebra, moving quickly through its Arab roots and European development and concentrating mostly on the 19th and early 20th centuries. Very accessible.

———. *Science Awakening I: Egyptian, Babylonian, and Greek Mathematics*. 5th ed. Dordrecht: Kluwer Academic Publishers, 1988. An excellent general overview of the mathematics of ancient Mesopotamia, Egypt, and Greece.

Varadarajan, V. S. *Mathematical World*. Vol. 12, *Algebra in Ancient and Modern Times*. Providence, RI: American Mathematical Society, 1998. Algebraic topics chosen from different historical periods. Particularly interesting for its treatment of the development of algebra in South Asia.

Whiteside, D. T. "Patterns of Mathematical Thought in the Later Seventeenth Century." *Archive for History of Exact Sciences* 1 (1960–1962): 13–388. Though published as an article, this is a really a book—the definitive description of how European scientists of the 17th century thought of the mathematics with which they were working.

Wigner, Eugene. "The Unreasonable Effectiveness of Mathematics in the Natural Sciences." *Communications on Pure and Applied Mathematics* 13 (1960): 1–14. The classic article by the Princeton physicist, in which he explains his views of mathematics and physics and gives examples of the power of mathematics to inform our understanding of the physical universe.

Notes